T0189164

IT PROJECT PROPOSALS
Writing to Win

Whether responding to a tender from a potential client or pitching a new IT project to the Board, a well-written proposal can be the difference between success and failure. *IT Project Proposals Writing to Win* can help you to create high quality, persuasive proposals that will stand out from the crowd. The author explains how to determine the reader's basis of decision and the writer's unique selling points. It discusses the structuring of documents, the secrets behind persuasive writing, and the basic grammar and punctuation rules that will prevent writers from destroying a good argument through bad presentation. Case studies and numerous examples show how the techniques described can be used in real-life situations. The book also introduces an automated questionnaire allowing any IT proposal to be reviewed and rated. It is invaluable for IT managers, consultants and for anyone else producing internal or commercial proposals promoting software products or services.

Paul Coombs is an expert in bidding, costing and proposal-writing for IT projects, following 25 years of experience with the international systems house Logica CMG and the media organisation, Reuters. He has worked on mission-critical developments for EMI, London Underground, IBM, the BBC, British Airways, and for a large number of major financial institutions. He has developed winning proposal strategies for large fixed-price projects in industry sectors as diverse as finance, defence, government, media and communications. He currently works as an independent consultant, running courses in estimation, project-writing and the bid process as well as undertaking cost/benefit analyses and proposals for specific projects. He is the author of *IT Project Estimation: A Practical Guide to the Costing of Software*.

IT PROJECT PROPOSALS
Writing to Win

PAUL COOMBS

CAMBRIDGE
UNIVERSITY PRESS

CAMBRIDGE
UNIVERSITY PRESS

University Printing House, Cambridge CB2 8BS, United Kingdom

Cambridge University Press is part of the University of Cambridge.

It furthers the University's mission by disseminating knowledge in the pursuit of education, learning and research at the highest international levels of excellence.

www.cambridge.org
Information on this title: www.cambridge.org/9780521612579

First published 2005

A catalogue record for this publication is available from the British Library

ISBN 978-0-521-61257-9 Paperback

To the staff of the Whittington Hospital, London

Contents

Microsoft Word

I have made some specific references to the facilities offered by Microsoft Word because it is the package that many people use to prepare their proposals. If you use a different word processor then substitute its equivalent operations or facilities. Microsoft Word is the registered trademark of Microsoft Corporation.

CHAPTER 1

Introduction

DOES GOOD WRITING MATTER?

All writing aims to achieve results. Whether the aim is to communicate information, to influence a decision or to stimulate the imagination, the language needs to be marshalled and controlled. There are many ways in which the desired aim can be achieved – and many more ways of getting it wrong.

The aim of *this* piece of writing is to improve the effectiveness of your proposals by making better use of language. By a "proposal", I mean any sort of document that is making a pitch: to your customers, to your manager, to your lover or to anyone whom you want to convince to take some action. We all receive many proposals, large and small, every day. But too often, although the information is accurate and the answers to our problems are in there, the way in which it is all presented is unconvincing and error-ridden. And so we don't buy; we don't take the action the writer wants us to. Maybe nothing could have been done – the ideas were rotten anyway – but possibly, if the writer could just have found the right words, we would have been convinced.

The purpose of this book is to help you find those "right words" for the proposals you produce, specifically those that seek to win or to initiate major IT projects. This is not to say that it is inapplicable to other kinds of proposals. If you work in another field then I'm sure you will find a great deal of useful advice – but most of the examples come from the world of IT. The book is not about sales techniques or specific technologies. I am going to assume that you have completed all the pre-paratory work and that you know what you want to say. What I am concerned about is that you express your brilliant solution or your revolutionary concepts in a way that is effective and persuasive.

You may believe that communication is all that matters and that your readers will forgive your inept English so long as your ideas are expressed somehow. I disagree,

1

for two reasons. Firstly, you will probably not have too much of your readers' time or attention. Every word must count, so your proposal must be organised and spun to achieve the maximum effect. Secondly, your readers may well be judging your errors in spelling and grammar as a reflection on the content of the text. If every document seen by your customers or by your managers has a few silly mistakes, what confidence will they have that you will fix every bug or that you can be trusted to work with a new client? If you can't be bothered to run a document through a spelling checker then will you take the trouble to create a solid and usable technical design?

A while ago, I was asked to review a proposal just about to be sent to an important potential customer. I turned to the first page and read the following:

Management Summary

Assuming the MSP Database and Infrastructure are in Place. The following deals with all Desktop and API's products which are currently available within MegaCorp Ltd. and try to find a suitable solution to enable these products Access and source their data from MSP. A migration period will be needed. Thus, how this migration will be reflected on current products and their impact on MegaCorps' customers, Identification of issues and the remedy for each of them Finally a plan reflecting a migration strategy will be proposed.

That management summary was never issued. But I suspect that many proposals just as incomprehensible have reached important customers, and that many reports have left their readers wondering about the competency of the individual or the organisation from which they were requested. A lot of effort wasted, just because the final presentation was not convincing or in a form that could be easily understood.

Usually a great deal of time and money is invested in the development of the technical solution to a customer's problem. Skilled, experienced people from many disciplines work together to design a complex, user-friendly system that meets all the requirements. And then what? Do we employ skilled, experienced people to communicate that solution? No, it is the technical experts who do that – and a right mess they usually make of it.

A proposal is your shop window – do you want to fill it with low-quality rubbish, or sparkling, attractive products? To win work through promoting your company's advantages, and to get decision-makers to undertake the actions you recommend, you must argue your case well and not undermine it with simple errors. Good communication is what will make or break the deal, so we need to dedicate as much time and as much skill to that as we devote to the technology. By using the language effectively, we can communicate better – and the better we communicate our ideas the more likely they are to be accepted.

FIGURE 1.1. Destruction of a writer's message

WHERE DOES IT ALL GO WRONG?

Figure 1.1 shows the five levels through which a writer's message becomes progressively destroyed. Each will be discussed in turn.

Inappropriate content

I have a problem: your task is to describe your solution to it. What do I want to see? That you understand the problem and that your solution is credible. I have a set of *rules*, either expressed or subconscious, that I am using to judge whether your proposal addresses my problem.

What is your aim? To make sure that I understand what makes your solution special, so I will choose *you* to solve my problem rather than any of your competitors. Thus, your proposal must satisfy my set of judging rules and express the reasons why your solution is the best.

What else should it contain? Nothing. There is no point in filling up your document with fascinating facts, however true they may be, unless they satisfy one of my judging rules or are relevant to your solution. So why is it that every proposal I read is full of blather, blarney, waffle, boilerplate, padding – call it what you will? I don't always want to be told about the history of my company – or yours. I don't want a chronicle of recent technical advances, I already know about recent trends in my sector, and I'm quite prepared to believe that you have a competent and experienced team without seeing a six-page résumé of each member. But – just to be annoying – sometimes I *do* demand these things. Your job is to find out what I want to see, what my evaluation rules are and what my problem *really* is. You must target every word you write specifically at these subjects; otherwise, you are wasting the time of both of us. Techniques for determining what should and should not be included in a proposal are discussed later in this book.

Confusing structure

The best way to wreck the effectiveness of a proposal is to pay no attention to its structure. Writers tend to throw down everything they know, in the order in which it occurs to them, without developing an argument or presenting a consistent case. This is particularly true of documents written by more than one person, where incompatible lumps of text are roughly stitched together, regardless of structure, repetition or sense.

There have been many times when I have reached some point in a proposal or technical document and realise that I no longer understand what I am reading or how it is connected to what has gone before. It is the same feeling that I get when watching a costume drama with a multiplicity of characters and wondering, "Which one is *she*?" Sometimes this is due to my own inattention, but more often the writer has failed to construct an argument sufficiently well, or has not given sufficient context to the point in the text where my comprehension has finally lapsed.

No matter how brilliant the solution that is being presented, you must ensure that the reader is led through it in a structured, logical order. The templates described in this book can act as the basis for proposals of all types and sizes.

Unconvincing prose

Dull, unpersuasive arguments are constructed through over-abstraction, evasion, repetition, lack of flow and overuse of clichés. Have a look at this example:

At present Management Information data is stored in a wide variety of different databases and the maintenance processes to collect and maintain the data are duplicated and inefficient. Several databases use outdated technology and consequently the applications to derive the data are also outdated. The aim of the new MIS Strategy is to unify together all these outdated databases into one central database. This will use up-to-date RDBMS technology which is maintained using a single data maintenance application.

If you can't see anything wrong with that, you definitely need to read on; this text is analysed and transformed later in this book. I will explain how to give your prose more sparkle and how to spin the words to maximise the effect you are trying to produce.

Poor grammar

Yes, our language is evolving. But it is not up to you to invent new grammar and punctuation rules. There's no need to be over-pedantic, but writing within the rules makes your arguments more persuasive, and avoids distracting the reader with puzzling or incorrect constructions. For example, I recently received an expensively produced glossy brochure from a software house, enticing me to use their products and services. A great deal of time and money had been devoted to its production, and I have no doubt that a team of creative masterminds had laboured for hours over the front-page slogan:

Hackitout Software – professionalism at it's best!

A single misplaced apostrophe and all that work was in vain. So learn the rules set out in this book, and don't let bad grammar expose you to ridicule or diminish the power of your prose.

Typos and spelling mistakes

Use of a word processor provides little excuse for spelling mistakes. But your document is not necessarily suitable for release just because the automated check shows no errors. This book provides some tips on how to review your work, and advice on how to eliminate the simple slips that devalue your message.

THE PROPOSAL LIFECYCLE

In a commercial context, writing the proposal is only a small part of the overall sales cycle. We also need to maintain contact with the customer, derive an appropriate solution, determine a price, analyse the risks, obtain approval, follow up the sale and

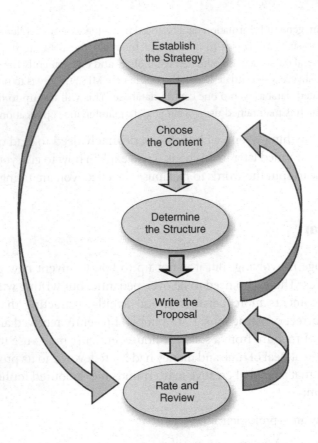

FIGURE 1.2. The proposal lifecycle

so on. But this book concentrates on the proposal itself. Figure 1.2 shows the process that will obtain a persuasive, effective result.

We will be returning to this diagram several times, but at this point you should notice that the process is iterative. We may need to revise the content and structure of the proposal as we write more of the text, and we will almost certainly need to rewrite some sections following internal or external reviews. And, during these reviews, we will be ensuring the proposal is tightly focussed with respect to the original strategy. The next chapter describes the first stage: how to determine that strategy.

CHAPTER 2

Establishing the strategy

THE ART OF PERSUASION

There are several methods to get decision-makers to do what we want. Coercion, blackmail, lies and flattery are some possibilities, although not necessarily the best ways to obtain the go-ahead for an IT project. Usually, we need to employ the most difficult method: persuasion. The decision-makers are not necessarily *against* what we are proposing, but there are other options open to them. So we must manipulate their thinking in such a way that they reach the conclusion that our proposal is the only way to go.

Arguments are rarely won purely by logic. Anybody, including your rivals, can analyse the problem and produce a solution. To *persuade* the decision-makers, you must present creative new ideas that are based on that underlying logic. Then you need to show an emotional commitment – maybe some excitement as to what can be achieved or a reassurance that your organisation is reliable and reputable. Because, once that emotional commitment has been communicated, you stand more chance of achieving empathy – the state where the decision-makers believe that you understand the problem, and can be depended upon to resolve it. And so you finally get to the point at which the decision is made. Something connects in the customer's mind, and persuasion is achieved. We cannot *guarantee* that will happen because there are too many variables, most of which are outside our control. But, unless we have built the foundations well, our proposal will never get to be the top choice.

There are several stages to the mental journey that the decision-makers must take before this happens – these are shown in Figure 2.1.

The first stage is to establish the common ground – to make sure that we have the same world-view and the same appreciation of the problem to be solved. As the Roman orator and statesman, Cicero, put it:

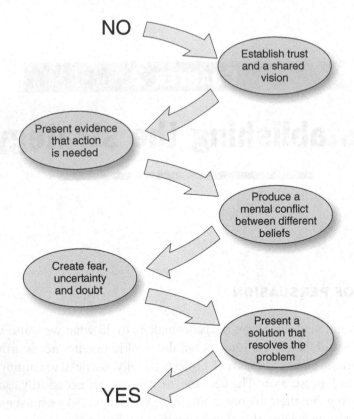

NO

Establish trust and a shared vision

Present evidence that action is needed

Produce a mental conflict between different beliefs

Create fear, uncertainty and doubt

Present a solution that resolves the problem

YES

FIGURE 2.1. The process of persuasion

> If you want to persuade me, you must think my thoughts, feel my feelings and speak my words.

It could be the technical aspects of the proposal that are of most interest to you because they provide a chance to use exciting new techniques that will look good on your CV. But such aspects may be of little interest to the decision-makers, who are more likely concerned with keeping costs down and sales up. You must see the world through their eyes and express your proposal as being the solution to a problem that they have. To do this, you must use their language – in particular, the exact words that they chose to express their problem – when discussing your solution.

The second stage is to present the evidence. This may include confirmation that there is a problem, proof that this problem is serious enough to require action, evidence that previous initiatives have failed, support from other sources and so on. The overall effect is to prove that the current situation is unsustainable – that "do nothing" is not an option.

The third stage involves creating some sort of stress. The decision-makers are now holding two conflicting beliefs: their current world-view and the evidence you have presented. For example:

- Your IT manager believes that the customer database is operating well, but you have just presented convincing proof that it is obsolete.
- Your customers thought the previous proposal that they read was the best possible response to their Invitation to Tender, but your solution seems better and cheaper.
- Your clients think their current IT consultants are doing a good job, but you have suggested some new ideas that those consultants would never have conceived.

It is uncomfortable to be hosting a mental battle between two different beliefs, so something has to give. The decision-makers may use your evidence to reinforce their existing views, they may reject your evidence or they may accept that their views need to change.

If they are now open to persuasion then we can build on this during the fourth stage by evoking fear, uncertainty and doubt: the FUD Factor. So we may assert that within the next few months the database will fill up, response times will deteriorate, hardware will need to be upgraded, the market will demand better products, new legislation will need to be addressed, costs will rise, the sky will fall – anything that will make ignoring our proposal seem a very bad idea. We want to evoke a slight feeling of panic – it is nearly too late, something must be done now.

And so the ground is laid for the fifth stage, where we present our solution and show the way forward. Our solution is so convincing and so appropriate that it restores the decision-maker's mental stability and leaves them feeling that we have the medicine to soothe their pains. Now the battle is nearly won – we have turned "no" into "yes".

When we review the persuasion process, we can see that emotions are important elements. Our proposal cannot be bland and neutral. It needs to show our emotional commitment and to invoke emotions in the reader – emotions like inspiration, encouragement, comfort, shock, stimulation and worry. To induce such emotions we must press the right buttons – the emotional triggers that will convince our readers that we are sincere, that we understand their problems and that we can be trusted to implement the correct solutions. To do this, we must ascertain two things: the conditions that will generate the desired emotions in our readers, and the particular elements of our proposal that will establish those conditions. The first of these is called the **Basis of Decision** and the second is termed our **Unique Selling Point**.

DETERMINING THE **BOD** AND **USP**

If you're experienced in selling, you're probably used to these terms. The Basis of Decision (BOD) is the set of rules the customer will use to decide which of a number of competing proposals is the best. The Unique Selling Point (USP) is the set of elements that we think makes our solution more appealing than anything our rivals can dream up. Ideally, the USP will successfully address each element of the BOD and our proposal will win.

Identification of the BOD and the USP is essential to the production of every type of proposal. Before you write anything at all, you must ask yourself:

- What criteria will my readers be using to decide whether my proposal is persuasive or not?
- What is going to be special about my proposal, such that it will have the best chance of achieving the desired effect?

Write these things down first, constantly turn back to them as the document is being written, and review the result with these elements in mind. The foundation of effective proposal writing is to make sure that *every word* appeals to the reader's Basis of Decision while stressing your Unique Selling Point. *All other words are redundant.* So how long should your proposal be? Exactly long enough to address the BOD, explain your USP and no more. Let's look at some examples:

1. *A technical report recommending a new hardware architecture*

   Here, the BOD would embrace the following factors:
   - Does the writer understand the current situation?
   - Do I understand what is being proposed?
   - Does it make sense?
   - Is it realistic?
   - Have the alternatives been considered?
   - Has everything I wanted to know been covered?
   - Are the costs reasonable?

   The USP may include:
   - Providing a context to the report.
   - Demonstrating technical competence.
   - Using the right terms (those actually employed by the reader).
   - Using diagrams, prose and layout to describe the proposed new architecture.
   - Providing facts – volumes, reliability statistics, prices etc.
   - Listing who was consulted, so the reader can see where omissions may have arisen or where particular ideas may have come from.
   - Describing alternatives and why they were rejected.
   - Stressing the benefits alongside the costs.

2. *An official application form for old people who may be eligible for government grants towards home heating*

    The BOD would include:
    - Do I understand this?
    - Does it apply to me?
    - Is there any danger or risk?
    - What do I have to do?

    The USP would comprise:
    - Making things very clear by using simple language, many diagrams and plenty of space.
    - Covering all possibilities, making sure the readers know what would happen once they apply.
    - Conveying a tone of reassurance without being patronising.
    - Stressing benefits rather than procedures.

3. *A user manual for a dealing-room system for bond traders*

    Here, the BOD would have to take into account that the traders are not going to want to reach for a manual during a busy day. Ideally, it will be read once and remembered, or consulted during quiet moments. When information is needed quickly, it must be easy to find.

    The USP may be to combine the functions of a quick reference with an in-depth description. This means splitting out the information likely to be needed in a hurry from wordier descriptions that the traders may read once or will only read when the pressure is off. The quick reference should not just include the commonly used functions, which will soon become familiar. It is the information that is used less often, maybe under exceptional circumstances, that will be needed quickly.

4. *A love-letter*

    It may seem calculating to determine a BOD and USP for such a purpose, but better that than leaving your partner feeling they are being addressed as "To whom it may concern..." My guess for the BOD is sincerity. If you are trying to persuade, cajole, or melt the hardest of hearts then establishing trust is the first step. Your USP will be the amount of your own personality that you can convey in the letter. One of the kindest compliments that can be paid to your informal writing is that the reader can imagine you there, speaking to them.

    I included this example to show up a significant difference between formal writing and personal communication. In business documents, you do *not* want your personality to shine through. The reader should get an impression of someone organised, clear-thinking and inventive, but not one of you as an indivi-dual. Your proposal represents your company, not yourself. This book is full of the quirks of my personality, but, if you were to read my business reports and proposals, you would find this tone has disappeared. Boring, but professional.

Looking at these four examples, you can see that some of the Unique Selling Points 'answer' the items in the Basis of Decision. This gives a clue that your proposal is heading along the right lines.

It is usually a great deal easier to identify the Basis of Decision than the Unique Selling Point. In commercial situations, you often have an Invitation to Tender or a Request for Proposal from which the decision criteria can be deduced. For internal projects, you know the current state of the organisation, so you can analyse the problems being faced by your managers. The only danger is in failing to put yourself in the shoes of the decision-makers – for instance, by stressing technological rather than business benefits. Some factors are almost always part of the BOD: your credibility as a supplier, your understanding of the background to the customer's business problem, your analysis of the specific problem, your attitude to the project's risks, your creativity, and your ability to express the solution.

In determining the Unique Selling Point, the emphasis has to be on the word "unique". For example, suppose you are bidding a fixed price for a system specified by the customer, in competition with other suppliers. Despite your best efforts, there is nothing particularly special about your technical solution; all of your competitors will be coming up with something pretty similar because of the nature of the problem. So you may just be competing on cost. Unfortunately, you know that your price is not going to be the cheapest. What, then, is your USP? Well, there has to be a reason for that uncompetitive price. Maybe your organisation has high standards, so is known for the quality of its work and for completing projects within their estimated time and budget. Those are your *unique* selling points over cheaper rivals. Hence, your proposal should not major on the technical solution but stress the quality and reliability of your service. Leave it to the competitors to talk up the marvels of the technology, which you can state as being a simple, standard problem. Remember, you need to establish a mental conflict in the decision-maker's mind that you can then resolve. So relocate that conflict to a different battlefield by stating that the decision should be made not on price, but on reliability and quality – the factors where you can prove your superiority.

Unsolicited proposals

So far, I have been assuming that you are producing a proposal in response to an Invitation to Tender, or a similar document, that the customer may have issued to many suppliers. Or else that you are making some sort of internal proposal within your organisation. However, there are other cases. For example, you may have explored some ideas during a meeting and the customer has asked you to put these in writing, maybe with some costs and plans. This is a tricky situation because without a written statement of requirements you may go a long way down the wrong path. If possible, you should make this a two-stage process. Firstly, write a document that states the problem to be solved and the requirements for a solution, effectively defining the BOD. Agree this with the customer. Then you can construct your proposal around this specification and stress your USP. There is a danger that the customer will decide to approach some of your rivals, using your 'proposal' as a specification, but at least you are in pole position.

Another possibility is that you have decided to issue a proposal without any request from the customer. For example, you may have a new product or service that you wish to promote. Here, too, it is a good discipline to spend some time listing what the customer's requirements and BOD *might* be. Your pitch can then be targeted to those specific needs.

In general, however, there is little if anything in the chapters that follow that cannot be applied to all types of proposal, solicited or unsolicited.

KNOWING THE READER

When determining the Basis of Decision, we need to identify the decision *makers* – usually those holding the cheque book. We must understand what will convince them, and anticipate the objections and questions they may raise. Their requirements may well differ from those expressed by technicians, who may have supplied the documentation to which you are responding, but aren't so concerned with the underlying business motives. It is very easy to create a proposal full of brilliant technical ideas, but which fails to explain how those ideas provide any actual benefit. We don't want the decision-makers to reach the end, think "So what?" and then toss our proposal on the slush pile. Consider this Management Summary, which begins:

> Hackitout Software is pleased to present its proposal to MegaCorp Ltd for the DSG/ PPSS. Our solution fully meets the requirements of PDS7763 V2 (G Richardson 23/7/ 97) and has been agreed with PSOS, PSEC and PSYS groups (See Appendix B sections B.2.3 – B.2.6). We recommend that the PPSS is migrated to the same platform as PDSS, and that the two systems become more closely integrated before migrating both to a new Sun/SOLARIS platform to meet the needs of DSG. To enable this we propose the use of the Java Cafe package that will enable a combined interface to be constructed from reusable beans and accessed through IE5.

Sometimes, this is exactly what the readers will expect; they know the context, understand the jargon and want the facts. But we would have to be very sure that this were the case before adopting such an approach. Do we really know the technical abilities of each member of the "management" who may read this summary? Will anybody be deterred from reading the rest of our proposal because it doesn't seem to address the truly important issues?

Often, the expected readers have many different roles, so some parts of the proposal may be judged using a different Basis of Decision from others. In such cases, it is important that each type of reader be directed to 'their' sections, not to arguments written to convince a different audience. Headings like *Management Summary* and *Technical Alternatives* achieve this, without any readers feeling that information is being withheld from them.

CHOOSING A STRATEGY

Once we have established the BOD and USP, and assessed the type of readers to whom our proposal must appeal, we can draw up the **proposal strategy**. This is not the same as a sales strategy, although it is a subset of that; it is the main theme of the proposal. A common mistake is to be "all things to all men" – your understanding of the customer's problem is unsurpassed; your solution is as innovative as it is appropriate; your prices are a bargain; your quality standards are incomparable; your employees are geniuses; your company is a shining example to the industry *and* your experience is unparalleled. It is unlikely that *all* these things are true; the customer knows that they will get what they pay for from software vendors just as they do from all their other suppliers. And, even if all these selling points *are* true, the resulting proposal will be unfocussed because it does not address the specific readership or their BOD. Readers may find some points persuasive, but these will be drowned by the long stretches of blather that they don't care about. We need to determine a simple, single strategy that will appeal to the specific decision-makers for *this problem*.

That strategy will be centred on a **major theme**. Although the proposal text may wander into all sorts of topics, it should always return to the major theme because at heart this is what we are selling. Typical generic themes are shown in Table 2.1; to these should be added any themes and USPs that are specific to your proposed project.

Often, one or two USPs don't seem to fit with the major theme. You must decide whether they should be rephrased or omitted so as to keep the proposal focussed. For example, let's say you have chosen cost as your major theme, but your product has a particularly good user interface that your rivals cannot match. You could write a section about the importance of the user interface, but this would be a distraction from the persuasiveness of your central argument. However, you can reconsider this USP in terms of your major theme. Your user interface *saves money* because errors are reduced,

TABLE 2.1. Generic themes	
Theme	**Typical USPs**
Cost	<ul><li>Price.</li><li>Return on investment.</li><li>Savings (in terms of money, time, resources, bandwidth etc.)</li></ul>
Technical	<ul><li>Innovative or advanced technology.</li><li>Superior specifications of chosen hardware or software.</li><li>Good fit of packaged solution.</li><li>Flexibility, performance, reliability, expandability etc.</li><li>Experience and skills of technical staff.</li></ul>
Quality	<ul><li>Well-defined quality procedures.</li><li>Reputation for quality.</li><li>Measurable and visible control over development processes.</li><li>Experience and skills of project management staff.</li></ul>
Credibility	<ul><li>Financial strength.</li><li>Company size, reputation and philosophy.</li><li>Experience with similar projects or customers.</li><li>Quality of line management.</li></ul>

operators are alerted more quickly, commands are processed more rapidly, fewer data entry clerks are needed or whatever. Now this USP can go into the list of reasons why your solution is the most cost-effective, and thus the proposal keeps its focus.

If you are in doubt about which strategy to choose then select the one that is the most positive. Negative strategies are the ones that play on fear, uncertainty and doubt: what will happen if the system is not introduced, or if a rival supplier is chosen. A positive strategy is based on benefit: all the good things that the customer will obtain once the system is running. Price alone is a rather negative strategy, which is why proposals that are fixated on costs are so rarely successful. If cost is your strategy then make sure that the *benefits* of your low price are stressed: payback time, return on investment and so on. A positive spin always reads better. The ideal is to make the customer feel that by working with you they will be moving in a definite direction, not just reacting to events.

As a final check on your chosen strategy, ask yourself two questions:

1. **Are we proposing what the customer needs or what we happen to have?** A packaged or recycled solution will sell – but only if you are able to express exactly why it is appropriate.
2. **Are we responding to customer needs or to the competition?** You must propose something *better*, not just provide a counter to all the features of your rivals' solutions.

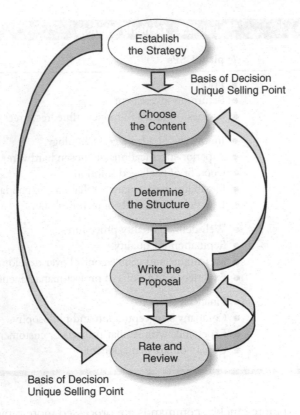

FIGURE 2.2. Proposal lifecycle, stage 1 complete

THE PROPOSAL LIFECYCLE – REVIEWED

Looking again at the proposal lifecycle in Figure 2.2, you can see that we have now completed the first stage. We are seeing the world from the reader's point of view, know their Basis of Decision, have determined our Unique Selling Point and have established a strategy.

During the review stage, we will assess the proposal with respect to the strategy, in particular to ensure that the completed document does not contain any irrelevant text. But now we can proceed to the next stage and determine the content of our proposal.

CHAPTER 3

Choosing the content

CONTENT AND STRUCTURE

Content is not the same as structure. We need to decide what we are going to say – the arguments that will win our case. And then we need to decide how that content should be presented, so the arguments appear in an ordered and persuasive sequence. As Figure 3.1 shows, the content is made up of a number of different pieces that are interlocked into the structure most suitable for a specific proposal.

It must be that way around: content then structure. We cannot start with a predefined plan and then try to force everything we want to say into that. This may seem self-evident, but many organisations mandate a fixed structure for all proposals.

Content

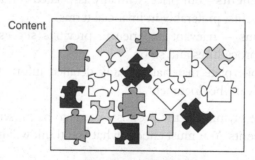

Structure

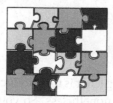

FIGURE 3.1. Content and structure

The authors feel they must adhere to this template, so some sections end up as "not applicable", or get filled with irrelevant blather. This is not writing that wins.

I am not going to suggest that there is some perfect combination of content and structure that will guarantee success for every proposal you submit. Instead, I will list some typical sections that a proposal may contain and describe the factors that need to be considered when creating each of these. You must decide which sections are applicable to your proposal and determine the specific content of each one. Having selected the most appropriate sections, you can then examine the plans in the next chapter in order to pick the structure that will best present the chosen content.

TYPICAL PROPOSAL SECTIONS

Regardless of what is being proposed, or the length of the proposal, the following elements are typically needed:

- **A Covering Letter** – obviously, this is unnecessary if the proposal is in the form of a letter, but in all other cases one should be provided.
- **A Management Summary** – for complex proposals, an additional Technical Summary may also be needed.
- **The Context** – the reason why the proposal is being made and a description of the current situation or problem.
- **The Technical Solution** – alternative solutions may also be presented and compared.
- **Project Management** – how we intend to achieve success, supported by organisation charts, schedules and so on.
- **The Costs and Benefits** – our price, with any associated conditions and assumptions, and the benefits, preferably in financial terms.
- **Our Qualifications** – relevant experience, previous successful projects, staff résumés, quality procedures and so on.
- **Appendices** – boilerplate material and any detailed information that does not belong in the body of the document.

Remember, I am not saying that the proposal needs to be presented in this order – although it has its merits. You must choose what you think will be required for each specific case.

The covering letter

The main purpose of a covering letter is to introduce the proposal. This may seem rather formal, but is good manners nonetheless. So you should state the proposal number or title at some point. As well as this, a covering letter may include the following:

- The context. It could be that your customer has been excitedly pacing the floor, doing nothing else while awaiting the arrival of your proposal. But it is more likely that it is just one of the many documents in their in-tray, and that they will need to switch into the correct mindset before they start to read.
- The key features of your proposal, particularly any aspects that will distinguish your solution from its rivals. Don't mention the price, unless it is *very* much lower than the reader will expect; you don't have enough space in a covering letter to balance that one cold fact against the merits of your solution.
- An explanation of why your company is the best choice.
- The validity period for any prices or schedules quoted in the proposal.
- Thanks to the customers for any help they have given.
- A request for the business.

The last point is the most important. Don't just end with some clichéd line like "If you have any questions, feel free to call". The point of a proposal is to get the reader to take the action you propose. The closure must tell them what they should do next: arrange a meeting, authorise your project to begin, sign a cheque or whatever. If this element is missing, you only do half the job – exciting the reader with your ideas, but failing to show the way to make those ideas happen. So the covering letter must end with an appeal to the reader to commit to your vision, through taking some action that you recommend. Make sure that the suggested commitment is limited and easy to under-take; if the initial step is too large, the decision-makers will be deterred.

The snag is that all of this must be done in a single page. Because you don't have many words to work with, you must eliminate the small talk that tends to appear in such letters. Remove "I would like to take this opportunity to thank you for the chance to present our proposal . . ." and get straight in with something like:

> Hackitout Software is proposing a complete turnkey solution for MegaCorp's new Customer Contact System. Our solution is . . .

The covering letter should be printed on your best, headed notepaper and addressed to whoever asked for the proposal. If possible, it should be signed by the main sales contact, not some Grand Panjandrum of a manager who has had no direct involve-ment with the customer. However, sometimes the letter acts as a formal acceptance of the requirements and an authorisation for the quoted price, so must be signed by someone qualified to do this.

The management summary

A rich source of examples for this book has been the "Management Summary" sections of proposals. These usually seem to have been written by someone who has not let their unfamiliarity with the document prevent them from believing that they can sprinkle it with fairy dust. The result is a summary created *by* management rather than *for* management, as we saw with this example earlier:

Management Summary

Assuming the MSP Database and Infrastructure are in Place. The following deals with all Desktop and API's products which are currently available within MegaCorp Ltd. and try to find a suitable solution to enable these products Access and source their data from MSP. A migration period will be needed. Thus, how this migration will be reflected on current products and their impact on MegaCorps' customers, Identification of issues and the remedy for each of them Finally a plan reflecting a migration strategy will be proposed.

The worst of this example's many faults is that it is not a "Management Summary" at all; it is a list of contents. A summary must embrace much more than that. It is the proposal in miniature – a précis aimed specifically at the upper-level, non-technical decision-maker. This is a difficult task in a page or two, but that's what's wanted.

But although the summary is more than a list of contents, we mustn't err in the other direction by including points that can't be found elsewhere in the proposal. The summary may use different language – maybe selling the business advantages rather than the technical detail of our solution – but a summary should be what it says, not a supplement to the text below.

Remember that the summary is the *only* part of your proposal that some decision-makers may read. Fine detail should be avoided. Instead, concentrate on what makes your proposal different, and on the features that will be compared with competing solutions. And make it specific to the system being proposed – no boilerplate text about the marvels of your organisation, its experience or its methods, unless this can be related to the problem in hand.

With these points in mind, it is clear that the summary cannot be tackled until the proposal itself is complete, or nearly so. The best method of creating a summary is to read through the proposal, copying its key sentences into a blank document. Arrange, join, simplify and condense these sentences into a coherent argument, making the summary into a mini-proposal that follows all the rules laid out in this book. The structure is likely to mirror that of the proposal itself. Make sure that you indicate that your solution will meet all the customer's objectives, and that you state exactly what you are offering. The result should be between one and four pages long, depending on the complexity of the system in question. With larger proposals, you may need to consider writing two summaries: a management one to encapsulate the business case and a technical one to provide an overview of the proposed solution.

The context

The "Context" section describes the current situation and the problem to be solved. The main difficulty is in avoiding the obvious – telling the reader too much that they already know, in particular by repeating information from the Invitation to Tender. We need to show not just that we have listened but that we have understood. Typical areas in which we need to demonstrate that understanding include the culture of the customer's organisation, their internal politics, trends in their industry sector, their competition, the overall aims of the proposed system, and any specific or unusual requirements.

What is needed is not a list of facts but a story – one with a happy ending when your solution is adopted. So don't start with an obvious, uninspiring line like:

> MegaCorp is the world's largest manufacturer of widgets, employing 2,365 staff, and with a turnover of $12.7m in 2004.

Instead, cut to the chase – the root causes of the problem to be solved.

> MegaCorp's market share in widget production has been declining for the past five years. One of the principal causes is that the AUTOMIS system, introduced in 1993, is not providing the information needed to manage a 21st century organisation.

There must be some reason why change is needed: internal initiatives, reorganisation, growth, dissatisfaction, technical advances, the competition, the economic environment, new legislation or whatever. Start with this, so you can then describe the problem that has resulted. This description can be followed with a history of any solutions previously attempted. In turn, this can lead into a discussion of the features that the customer will be looking for in the new system – the technical elements of the BOD.

Some of your assessment will not be news to the reader, but other parts will be an introduction to the way you are tackling the challenge. Be careful how you mix these elements together. It is tempting to blend some incontrovertible facts with some more contentious opinions in order to steer an argument towards your point of view. But if the reader disagrees with the opinions, or sees through this trick, they may feel misled. Signal your analysis with "in our view", "we consider" or similar phrases. And don't ridicule any solutions previously attempted; it could well be that the people evaluating your proposal either designed those solutions or commissioned someone to do so.

Our qualifications

Again, it is important in the "Qualifications" section that the reader is not given too much information that they already know, or don't care about. Firstly, you must consider how much work you have undertaken for the customer before; if you have a long-standing relationship, there is no need to include the high-level introductory material. Secondly, consider the relevance of the information with respect to the BOD and USP. If you think that the size, stability, experience or quality offered by your organisation are hot buttons then you need to include the relevant evidence. But it must *be* relevant. Don't include long stretches of boilerplate text that may or may not be applicable to the project being discussed, and don't hide all the evidence away in unreferenced appendices.

Typical contents of the "Qualifications" section include:

- Corporate profiles and financial results.
- Job résumés from previous projects you have undertaken.
- Staff résumés.
- Quality methodologies.
- Reprints of articles about your organisation or its products.
- Industry surveys.
- Product data sheets.

The technical solution

Given that *all* of a document should be targeted to meet the Basis of Decision and display your Unique Selling Point, is there any use for text like this?

> One approach would be to develop the MDIS format in the context of the Electronic Industries Association's CASE Data Interchange Format (CDIF) standards. The CDIF family of standards is "primarily a description of a mechanism for transferring information between CASE tools" and supports multiple semantic layers and transfer formats. The current version of the CDIF standards represents a multi-year effort, expected over time to be adopted as an ISO and ANSI standard. To this end, the goal of the CDIF standard is to be as semantically complete as possible. However, because what constitutes metadata evolves as various types of software technology are developed, the EIA has established an extensible standard and encourages the development of working groups to address new areas of interest. Adopting this approach carries with it two obligations: the Metadata Coalition must appoint one or more members to track the CDIF standards; and every vendor supporting the MDIS format must subscribe to CDIF publications to avoid violating the EIA's copyright on those standards.

Yes, such text does have its place. Our readers will want to feel that we understand what we are talking about – that we have a good grasp of the job to be done and the technology to undertake it. This is certainly going to be a part of their Basis of Decision. The best way to demonstrate our understanding is to use crisp, authoritative prose – as in the example. The technical decision-makers will be familiar with the jargon and will realise that we know our stuff.

But you must signal that such sections are targeted at those readers who will understand them. More importantly, you must be aware that the aim is not to show off, or to shove in all the information that you happen to know. Remember that you are writing a proposal, not defining the complete solution. You want to provide just enough to prove your expertise, implying that you could go into more detail if you wanted to. Technical competence is only one factor – one that must be balanced against the other aspects of your USP. If 90% of the proposal consists of technical explanations, however brilliant, then that balance has been lost.

Project management

Our customers may have been impressed by the ingenuity and elegance of our technical solution, but we also need to prove that we can implement that vision. The "Project Management" section deals with the specifics of this project, whereas general points about management techniques and quality procedures are covered in the "Qualifications" section. Typical contents include:

- The overall management approach.
- The project organisation – staff roles and responsibilities.
- How the relationship with the customer will be managed (i.e., who in your organisation will deal with customer contacts in each relevant area – management, technical, operational, administrative and so on).
- Change control procedures.
- How relationships with any subcontractors will be managed.
- Implementation schedules and milestones.
- Testing procedures.
- Acceptance procedures.
- Installation and training schedules.
- Maintenance and warranty arrangements.

Such information need not be purely factual. You can add 'selling' text, so long as it is positive, relevant and interesting. For example, you may show how the project organisation you have chosen makes success more likely, or how the schedule meets the customer's requirements without being over-ambitious.

The costs and benefits

In most proposals, you will need a section dealing with prices or costs. Some say it should be at the start because most decision-makers won't be interested in wading through reams of technical detail before finding the one thing that interests them: the price. Others are of the view that no costs should be mentioned until we have had a chance to promote what is being proposed. You will have to decide where the balance lies – how important is the price in relation to other aspects of the BOD, and how big a factor is it in your USP?

Typical contents of the "Costs and Benefits" section include:

- A breakdown of the cost – for example, by phase, function or site.
- A list of deliverables.
- A kit list for hardware items.
- A list of assumptions.
- An analysis of risks.
- The proposed invoicing schedule.
- Commercial terms and conditions.
- Warranty conditions.

Analysis and quantification of many of these areas involves the creation of a **Cost Model**. This should include a list of estimatable tasks, the project plan, a risk analysis, staff costs, capital costs, ongoing costs, cashflow and price breakdowns. My book *IT Project Estimation: A Practical Guide to the Costing of Software* describes how a model for any specific project can be created and used.

One of the purposes of the "Costs and Benefits" section of the proposal is to define the scope of our offer to the customer. Providing a list of deliverables and a kit list should be sufficient in most cases. However, if there is any possibility of ambiguity then append a list of items that will *not* be supplied. You should not be trying to hide such things; your credibility will be fatally undermined if the customer spots any attempts at deception.

There is nothing wrong with stating your assumptions; they show that you have fully analysed the problem. However, listing them can be a difficult balancing act. If you include too many, it looks like you are trying to weasel out of commitment. But shortening the list may increase the risk. The key is to distinguish between those factors that are within your remit and those that are the responsibility of the customer. The former should be built into your price, by means of a contingency allowance. The latter are best presented not as a list of assumptions but as a list of deliverables from the customer to you – a counterpart to the inventory of items you will be supplying to them. The customer then understands their obligations, and the risks to which they are a party.

Don't hide costs. There are a zillion ways to do so – and your customer will have seen them all. The most common method is cited by Scott Adams in his book *Dilbert and the Way of the Weasel*:

> When you give a price quote, leave out as many costs as possible. This will make the price seem too low to resist. Most customers will forget to ask about things like taxes,

installation, delivery fees, insurance, warranty extensions, service contracts, cables and whatnot. A skilled weasel can convince the average ignorant customer that a $1000 item actually sells for 5¢.

Which would you rather have: an honest, comprehensive list of prices that pulls no punches, or a sneaky list that you have to question and interpret in order to get the true cost? Let the competition play those games.

A proposal should not just concentrate on the costs of the proposed project; it must also analyse the benefits. Many proposals fail to mention the advantages of implementing the work described. The writers assume that the customer knows what these are, so they concentrate on new information, like the costs. However, this unbalances the argument. We must dangle those benefits in front of the readers and convince them that they are well worth the price. At the least, re-listing the benefits will demonstrate that we understand that the project is in the customer's interest, not just a way to keep us in beers. Better still, we should try to add value to the benefits that the customer already recognises. Perhaps we can find some more – maybe some indirect advantages that they may not have considered. Or possibly we can cost all of the benefits and then demonstrate that the project price will be reclaimed at some point in the near future. Advanced techniques for determining the return on investment are beyond the scope of this book, but even a simple diagram, showing when the benefits will be realised and when the costs will be recovered, is more than many proposals provide. If you wish to read more about cost/benefit analysis, rate of return, break-even points and all the other factors that "make the numbers sing to management", I recommend *Making the Software Business Case* by Donald J. Reifer.

The appendices

Appendices fall into two groups:

1. Specific information that is too detailed or too unwieldy to be included in the body of the proposal. For example, there may be a Gantt Chart of the delivery schedule, or a table of part numbers for the equipment that will be purchased. In general, anything that is long and dull, which will interest a few readers but bore the rest, is a candidate for an appendix.
2. Generic information, usually pre-printed. This may include:
   - Commercial terms and conditions.
   - Staff résumés.
   - Quality methodologies.
   - Job résumés from previous projects you have undertaken.

- Environmental polices.
- Regulatory statements.
- Non-disclosure agreements.
- Draft contracts.
- Corporate profiles and financial results.
- Reprints of articles about your organisation or its products.
- Industry surveys.
- Product data sheets.

All appendices must be referenced within the body of the proposal, not just shoved in because you have the information and you think that someone might be interested. In particular, staff and job résumés should be targeted towards the skills and experience needed for this particular proposal, not just some boilerplate text chosen from a library.

Appendices should be numbered in the order in which they are referenced. The "Appendices" section itself needs an introduction and its own list of contents. The introduction is just something like this:

> This section provides additional information about Hackitout Software, its products and services, and how they relate to the XYZ system being proposed for BigCo. There are references throughout our proposal to the material to be found in this section. The appended information comprises . . .

THE PROPOSAL LIFECYCLE – REVIEWED

We have now established the message that we want to send to the reader and have assembled the evidence that will support that message. Hence, we have completed the second stage of the proposal lifecycle, as shown by Figure 3.2.

We have not written anything yet. Before that, we must decide what we want to say, pick an appropriate plan and expand it down to the level of sections and paragraph headings. *Then* we can start to write. Maybe you would like to proceed in a 'top-down' manner, expanding each of the section headings with notes summarising what needs to be said, and then building these into complete sets of paragraphs. Or maybe you would prefer to complete the document section by section. Either way, you must be prepared to erase, rewrite and restructure as you proceed. Some people are afraid to axe existing text – particularly if someone else has written it – in case some nugget of an idea is lost. So they force the unallocated prose into inappropriate places, destroying the flow and hindering the development of the

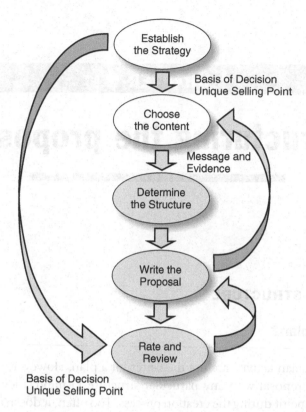

FIGURE 3.2. Proposal lifecycle, stage 2 complete

arguments. If you decide that some point is just too good to miss, you must alter the structure to accommodate it. But remember the rule: *every word* must appeal to the reader's Basis of Decision, stress your Unique Selling Point, or both. Everything else is redundant.

CHAPTER 4

CHAPTER 4

Structuring the proposal

DOCUMENT STRUCTURE

Why have a plan?

Few fields of human activity are not the better for a plan. However, most people do not start on a proposal with any particular structure in mind; they hope one will emerge at some point during the creation process. Too often, it doesn't, and the result is a hotchpotch of ideas with no theme or conclusions. We can't just throw down everything we know, with every alternative that we have dreamed up, in the hope that some of it will strike a chord with the reader. Examine this next example. The writer has some revolutionary, exciting, even rather crazy ideas. But he is trying to appeal to his managers, the people who will change the direction of the whole company to embrace those ideas. What will they make of this?

The methodology and principal is based on natural laws of new physics where we will attempt to create a phase transition at a lower level in order to allow the objects self organise into like structures. The simplest analogy of this process by which we will recompile the data is based on theories of Cellular Automata developed originally by Von Neuman (founder of Game theory) and Ted Codd (the inventor of relational database). The theory was expanded by Langton at the Santa Fe institute with their research into "artificial life". The "game of life" as it was termed looks similar to a game of "Go" where you have white and black pieces having negative and positive values that represent alive or dead states. Thus through a modified "Game of Life" we could eventually completely self organise our databases and create a visual discovery tool. Once the program is set within the closed systems universe the pieces begin to interact with each other much like Go with pieces living, and dying expanding and contracting eventually beginning to achieve a critical mass of interaction based on non-linear fluid dynamic flowing and forming and reforming. Eventually I believe that repetative higher state structures will begin to form within the closed universe. The structures and movement is representative of higher level complexity formed out of lower level chaos through infusing the system with enough energy to create a phase transition thereby creating an emergent quality and structure. Thus it will be with our databank oblets.

I could ask for no finer example of the "just throw it all down" method. It makes no logical sense (I particularly like the "thus" about halfway in), it assumes the reader has all sorts of knowledge about terms and concepts, and the grammar is appalling. But there are some brilliant ideas in there (trust me on this). To communicate these ideas, the author needed to sort out what he so rightly calls the "lower level chaos", and set his whole argument within a logical, planned structure.

If several different people are writing the proposal, defining an initial plan is even more important. Someone must take overall control of the document structure, and be responsible for standardising terms and conventions. Only once everyone understands exactly what is needed for their section can they begin to write. If you just let everyone loose, hoping to cobble together the resulting text into a single, well-structured, flowing, convincing proposal "at the end", you will inevitably be disappointed. Also, once you have a plan you can see how much progress you have made and how much remains to be written; again, this is an important consideration when the proposal has several authors.

Scientific versus business document structure

A typical scientific paper presents all the evidence first, leading up to a conclusion that, by the time you have reached it, seems inevitable and correct. Many writers learned this format at college and are convinced that it is applicable under all circumstances. But most readers will not read your entire proposal, so you have to get your conclusions and recommendations in front of them early. If you start with the nitty-gritty detail, you will lose their attention. Managers want to see the *answers*, not the list of problems or our assessment of the current situation. The last thing we want is for the reader to struggle through fifty pages while thinking, "I *know* all this". If you have confidence in your recommendations, put them up front so the reader immediately realises they are holding something relevant and useful. If they don't want to read the whole thing then that's fine – we've maximised the use of the attention span we were allocated. So, as Samuel Goldwyn once instructed the director of a new film:

Start with an earthquake and build up from there.

Announcing your plan

Whatever the plan for your document, you must let the reader know what it is. Later, I will discuss how to make the Table of Contents into a mini-summary of the whole document, as well as revealing its structure. But, because many readers skip the Table of Contents, there is also no harm in having an introductory paragraph such as this:

> In this proposal, Section 1 describes our understanding of the current situation; Section 2 defines our commercial terms for the implementation ...

This is often best set out as a bulleted list. Such a list can also be used to declare the structure of each individual chapter or section.

Naturally, having declared your plan you must stick to it. For example, if you have a section containing all the costs then don't quote additional prices in other sections.

TYPES OF PLAN

In the following sub-sections, I will discuss some possible plans for proposals, and the circumstances under which they might be adopted. Of course, you do not have to adhere to any of them; often a combination of approaches is appropriate. Certainly, no template should be so rigid as to prevent modification to suit the requirements of a particular project or readership. So these models are designed to inspire, not to provide a straitjacket. Just remember that you always need *something* sketched out before you start to write.

Point by point through the BOD

Having identified the reader's Basis of Decision, the proposal can cover each point in turn, starting with the most important. The whole thing may be preceded by some context and finished with a suggestion as to what should happen next.

A disadvantage of this plan is that you have to be sure that the Basis of Decision is exactly right. Otherwise, there is a risk that the reader will begin to think, "So what?" or "What about the *real* issues?" So this approach is most suitable when you have been able to agree the Basis of Decision with the customer, or when they have told you exactly what they are looking for. For example:

> You asked us to recommend a system that is expandable, flexible and reliable. Hence, we are recommending the Babbage 7645. It is:
>
> - Expandable because ...
> - Flexible because ...

- Reliable because ...

Hence, the Babbage 7645 meets your criteria for the proposed application and is our recommended solution.

Point by point through your USP

This reverses the previous approach. You go through the key points of your solution, showing that they address each element of the Basis of Decision. For example:

Our recommended solution is a combination of your existing PCs running standard browser software at the front end, with a new Babbage 7645 to support the BigCo Database and run the overnight batch processes. The rationale behind this is as follows:

- Using the existing PCs will reduce costs because there will be no need to purchase 100 new workstations or to upgrade the existing ones. Reliability will not be compromised because the existing PCs are relatively new, and can always be replaced by industry-standard products if they fail.
- A standard browser is a cheap solution for the front end because the product is inexpensive and because development times for the screens will be reduced. Such products have proved their reliability ...
- The Babbage 7645 model is a highly cost-effective ... reliable ...
- BigCo database ... cheap ... reliable ...

Hence, our recommendation meets your stated need for a solution that will provide high reliability for the minimum cost.

Problem-cause-solution

This model is ideal for a short report or a management summary. The *Problem* is why you were asked to write the proposal in the first place; it allows you to mention the Basis of Decision. The *Cause* is the result of your analysis and proves that you understand the problem. The *Solution* allows the Unique Selling Point to be stressed.

Two things are important with this model. Firstly, you must not jumble the problems and solutions throughout the proposal; all your text must be ordered and structured to fit into the correct section. Secondly, you must ensure that the causes and the solutions can be related back to the problems. There will be some overlaps between the three sections, but it is important that the reader be able to see exactly what applies where. One way out is to number the problems, referring to these numbers in the remaining two sections. Alternatively, give each problem a mini-title such as "the capacity limitations" and then refer to that.

Logical order

This applies in the following situations:

- You have been given a list of questions to be answered.
- You are describing a series of events that will happen (or have happened) over time.
- You are describing a process from start to end.

In such cases, it is self-evident how the text should be ordered. If there is a choice, start with general points and then move to the specific, or start with simple points before tackling the complex.

Predefined order

If your customer has provided a format for your proposal then you must follow it in the best way that you are able. Failure to do so may give the impression that you cannot meet even the simplest of requirements. Worse still, neglecting the given format will sometimes result in your proposal not being considered at all.

Unfortunately, you will find that there are unmissable points that do not fit into the given format. Worse still, the format may distort your arguments so you can't address the Basis of Decision or stress your Unique Selling Point effectively. There are several ways around this, such as adding extra sections to the given structure, preceding the whole document with a summary, or adding an appendix. But you must ask yourself whether such additional information really is of interest. For example, it is quite common for a customer's suggested format to omit the sections that would normally be filled with boilerplate text about your company's history and experience. The customer is signalling that they want the facts, not the waffle. It will not impress them if you ignore this hint and provide the waffle anyway. You certainly cannot fill the predefined sections with "see attached document" and then append something structured *your* way. Nor can you omit sections that may seem awkward or irrelevant. If you are given a proposal format then one element of the customer's Basis of Decision will be the extent to which you adhere to it.

Questions and answers

This structure consists of a set of questions that may be troubling your potential reader, together with your solution to each. The format is ideal for short proposals to senior managers with the attention span of a hyperactive butterfly. The style should be short and simple, for example:

What is the recommended approach?

A new system to act as a staging post for the maintenance of data before it is released to the data warehouse.

Will this result in two copies of all the data?

No, the data will be cleared from the data maintenance system once it has passed all quality checks.

Why can't the data warehouse itself be used for data maintenance?

The warehouse system is not efficient when "write" operations are applied at the same time as multiple "read" operations by the end-users. Response times, already slow, would be further reduced.

Ideally, you should present the questions in the same order as the reader would think of asking them, while still covering the entire solution.

Scrap

This structure, and SOAP, the next one, are described in Martin Cutts' book, *The Plain English Guide*. SCRAP stands for:

- Situation
- Complication
- Resolution
- Action
- Politeness

Such a structure is ideal for shorter proposals: letters rather than documents. The *Situation* describes the business problem to be solved, establishing the Basis of Decision. The *Complication* introduces any subsidiary factors, such as why an obvious solution cannot be employed. The *Resolution* lists what we recommend and is where our Unique Selling Point can be stressed. The *Action* describes what the reader should do now, if the benefits of the solution are to be realised. Finally, the *Politeness* allows a closing remark thanking the reader for their attention, acknowledging help received, or anticipating a meeting to discuss things further.

Soap

This format is very similar to SCRAP but is more amenable to longer proposal documents. SOAP stands for:

- Situation
- Objective
- Appraisal
- Proposal

The *Situation* is a description of the current environment; it should be factual and not biased towards any particular solution. The *Objective* is an assessment of the business aims of any proposed changes – essentially a statement of the Basis of Decision. The *Appraisal* allows the various alternatives to be considered and leads naturally into the *Proposal*, where the winning alternative can be considered in more detail.

Requirements-solution-benefits-costs-proof

Winning Business Proposals by Deiric McCann cites this format as "the universal winning proposal model". It can be scaled up, even to the largest projects. The plan divides the proposal into five sections:

- **Requirements** – the problem the proposal is trying to solve, which will imply most of the Basis of Decision.
- **Solution** – technical details of your solution.
- **Benefits** – showing how the Basis of Decision is addressed by your Unique Selling Point.
- **Costs** – all commercial terms and conditions.
- **Proof** – justification of what is said above, which is often in the form of appendices listing your capability and experience.

Illustrations first

If each picture is worth a thousand words, there seems a good case for presenting these first, then basing any remaining words around them. For example, if the proposal is a design for a new hardware architecture, a diagram showing all the boxes and interconnections will be the centrepiece. I have seen proposals where such a diagram appears several times in different guises, to show firstly the topography, then some key data flows, then how resilience is achieved, and finally the cost of each element. So, if your proposal can be expressed in a few simple pictures, it will be effective to get these in front of the reader early on, backing them up with progressively more detailed explanation.

Summary first

This is effective in cases where a sales person is guiding the 'politics' of the approach, or where a senior designer is specifying the details of the technical solution. That person writes the Management Summary and then the proposal team develops the remainder of the document in more detail. There are two risks with this approach. Firstly, the summary must encapsulate everything the proposal will say, *before* it is written. Secondly, the team must be trusted to conform to the template laid out by the summary. However, there are often cases where the key person cannot provide sufficient time to write the entire proposal, so this approach must be adopted.

Do-it-yourself

Just draw up a list of headings for all the points you want to make or all the topics you wish to discuss. Type them into the word processor in any order. Now try to classify and structure the list into some sort of framework. I created this book like that – developing some individual points almost to their final form, dotting about here and there as inspiration struck. I left some empty sections as placeholders, such as one titled *Page Layout*, which I moved around until I was ready to write them or (as in that particular case) until deciding to leave them out. New ideas emerged as the book developed, sometimes implying a complete revision of the original plan. But there was always a plan.

SECTIONS AND PARAGRAPHS

Now we have determined the overall framework, we need to consider how our message is going to be presented throughout the sections and paragraphs within that structure. You don't have to construct your proposal in a completely top-down manner, but it will certainly help if you have an idea about what you are going to say within each section before you start to write it. This is particularly true if the proposal is being written by a team because specific sections can then be assigned to the most suitable authors.

Size and order

A single paragraph or section is not a dumping ground for everything you know about a topic. Each section, sub-section and paragraph should have some sort of point to make – the smaller units contributing to the larger arguments. So you must be prepared to wield the axe. Decide on your line of reasoning, boot out anything irrelevant to that and make sure each sentence makes a specific contribution to your argument.

If you are ever troubled about when to start a new paragraph then the rule is quite simple: stop when you have made your point. It doesn't matter if this takes one line or one page, although it would have to be a very abstruse point in the second case. Typically, a paragraph will consist of three to six sentences, and a sub-section of between three and six paragraphs. Less begins to look a bit scrappy: more and the reader starts to lose the trail.

Good rhythm makes writing more appealing. It is obtained by varying the sentence and paragraph lengths while making the argument in the shortest number of words. There is nothing wrong with very short, even single-sentence, paragraphs. Used sparingly, they can make a point very effectively.

A good tip is to imagine that every paragraph has its own heading. Look at the list of these headings and decide if they are well ordered, flowing into each other to create a convincing line of reasoning.

Structuring paragraphs as triangles

The ideal paragraph can be considered as a triangle. The first sentence summarises the whole; subsequent sentences support and expand on this theme. As the paragraph progresses, the argument is developed by means of illustrations, examples or additional evidence. Look at the extract from Darwin's *Origin of Species* shown in Figure 4.1. Admire the flow from the initial theme (mind-boggling in its day) through to the three examples, which get progressively more complex and specific.

The triangular structure helps readers quickly identify arguments that they don't want to pursue and hence to speed-read the document. For it is a sad fact that most readers will treat reading your prose – jewel-like as it may be – as a chore rather than a pleasure. They will want to fillet out the specific points of interest to them. When they find something interesting (as we must hope they do), the document and paragraph structure will help them to explore further.

If necessary, the triangle to develop an argument can be preceded by a sentence to set the scene, or it can be followed by a short conclusion, or both. But it is important that any additional words be relevant to the central theme and do not ramble off into other areas, however interesting. Such points should have their own paragraphs.

If you can, make the sentences containing your central themes positive, novel, radical, surprising, controversial and shocking. They are the peppers in the stew.

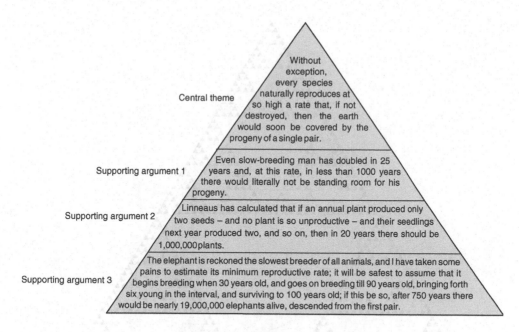

FIGURE 4.1. A paragraph triangle

Sir Arthur Eddington begins a chapter in his book *The Philosophy of Science* with the captivating sentence:

> I believe there are 15,746,724,136,275,002,577,605,653,961,181,555,468,044,717, 914,527,116,709,366,231,425,076,185,631,031,296 protons in the universe, and the same number of electrons.

Immediately, you wonder how he can be so definitive, so already he has drawn you into the paragraph that follows.

As for paragraphs, so for each section and the document as a whole. A section starts by summarising its major point, which is then expanded through a number of paragraphs to develop that theme, which in turn may be followed by a concluding thought or summary. So your proposal will consist of a hierarchy of triangles, whose tops represent the decomposition of the main theme into progressively smaller supporting arguments. This can be illustrated by the beautiful fractal figure known as Sierpinski's Gasket, shown in Figure 4.2.

Flowing your arguments

Successive paragraphs must link together smoothly. The effect should be a series of hammer-blows to bash home the theme of that section of the document. As the

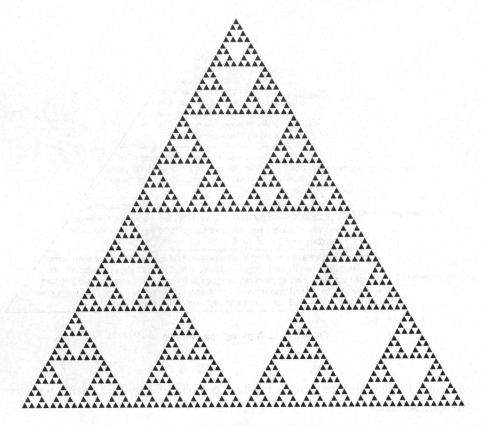

FIGURE 4.2. Sierpinski's Gasket

argument develops, you must keep the reader aware of where they are by signposting the major points as they are reached. However, you must make sure that you don't forget the signposts after the first couple of points, leaving the readers lost in a maze of unconnected arguments.

Flow is important within individual paragraphs too; the following example is just a collection of unconnected sentences and of clauses weakly linked by *and*, making the flow halting and clumsy:

> The weekly development meeting was held on Thursday, and most of the development managers attended. Fred Smith described the new architectural vision. The proposed requirements gathering process was reviewed. The team structure for the analytics team was drawn up by Joe Brown. The architecture was well received and Fred is travelling to the US next week. The new process was agreed to need more buy-in from the product groups. The usefulness of such meetings has been generally agreed to be positive and a weekly meeting seems about the right frequency. The next meeting will be on Wednesday 11th and hopefully Ian Jones will be there.

There are many ways of linking paragraphs and sentences together smoothly – some are listed below. You should aim to employ a good mixture of these methods, so your arguments do not become too predictable and stodgy.

- An announced structure that tells your reader what arguments to expect. For example, you might want to stress that your solution is cheap, simple and reliable. These would make good themes for three paragraphs, each introduced with text like:

  Turning to the *simplicity* of our solution, we have decided to remove the ...

- A bulleted list that can introduce some alternatives, and be followed by a linking phrase, such as "All three of these alternatives will now be evaluated further".
- An index to each point, for instance by using *firstly*, *secondly* and so on.
- An introductory sentence at the end of a paragraph, such as:

  Such are the advantages of the proposed solution. But there are three drawbacks, as follows.

- A summary of the previous argument, followed by a new argument that builds on that, or explaining an alternative. The following paragraph might start with some phrase like "despite these impressive results" or "although these problems are serious".
- A hypothetical question, such as "But why is this?" and "How can this be achieved?"
- The repetition of a concept or a keyword, as here:

  Each local plant will be equipped with a Process Data System (PDS), enabling local collection of all types of data from the production systems. Data is sent from the PDS to one of the Central Repositories. Repository batch processes generate nightly reports that are ...

- A restatement of the main theme, followed by a new argument that supports it, for example:

  Another way to improve the system performance is to increase the system memory. An additional ...

- A pronoun, such as *it, this* or *they*, that can refer back to terms used previously. Care must be taken that the reference is not ambiguous, not to a previously unmentioned item and not to a term mentioned too long ago, as here:

  The argument for renewal of the database is flawed. The BigCo system has many features, but experience from the London office shows that it is not one that will hold water.

- A transitional word or phrase, which may be used as follows:

- To add a new point (*also, furthermore, in addition, moreover*).
- To make a comparison (*likewise, similarly*).
- To make a contrast (*however, conversely, nevertheless, on the other hand*).
- To introduce an illustration (*for example, for instance*).
- To remind the reader about a previous point (*as we have seen*).
- To restate something (*in brief, in other words, in short*).
- To indicate a chronological sequence (*earlier, in the meantime, meanwhile, next, then, to begin with, while*).
- To make a summary (*in conclusion, on the whole, on balance, to sum up*).

If these techniques cannot make your concepts flow freely, you probably shouldn't be lumping them all into one paragraph or section. When you have a new idea to present, you want to signal this to the reader by introducing some white space, making sure that the succeeding text develops the overall theme.

HEADINGS

In a newspaper, the headline is the story, reduced to a few words. Downmarket papers make their headlines excessively snappy, exaggerating the interest of the story below, but technical writers err in the other direction, giving individual sections a bland title such as *Commercial* or *Conclusions*. These just define the topic being covered without revealing anything about what is being said. But by making your section headings into 'headlines' – summarising the text below – you facilitate speed-reading and turn your Table of Contents into a summary of the whole document. For example, if a section deals with the relative merits of different printer manufacturers, don't title it *Printers* or *Choice of Printer*. Better would be *Advantages of BigCo Printers* or *Why We Should Use MegaCorp Printers*. The readers then know the conclusion and can choose to read the supporting arguments if they wish. If you can put sales points into the titles then so much the better – for example, *How Hackitout Ensures the Delivery of High-Quality Systems*, not *Quality Procedures*.

Sometimes it is easier to write the section under a bland heading, such as *Technical Decision*. Once the section is complete, you can pick out the main point and ensure that this becomes your heading.

Ideally, the heading will be a concentrated version of the first, keynote sentence in the ensuing paragraph. It doesn't matter if some of the words are the same; the title is not part of the text but a signpost showing where a particular argument or discussion can be found. Indeed, the text shouldn't flow directly from the heading, as it does here:

2.13 Printing Management Reports
This can be done by selecting the "Reports" option from the main menu ...

All the headings within a section should have the same grammatical form and read in a logical order – for example, "Entering Customer Orders", "Viewing Orders in the Warehouse", "Printing the Invoices".

Heading levels

One signal that a document's structure is wrong is an excess of sub-heading levels. There are two rules:

1. Each page should have at least one sub-heading or diagram on it, otherwise the text starts to look boring and skippable.
2. There should never be more than four levels of sub-heading. And don't cheat by adding unnumbered sub-sections, unless this is mandated by a predefined structure.

If you find you cannot adhere to these two rules, it is time for a rethink of the structure. This is usually achieved by inserting another major section and reassigning the sub-sections you have already written.

THE PROPOSAL LIFECYCLE – REVIEWED

We have now completed one more stage of the lifecycle, as shown in Figure 4.3. We now have a plan for our proposal, with all the section and paragraph headings in place, ready to be expanded with convincing text. As you write this text, you will discover further points that you need to mention, which may lead to a revision – large or small – of the structure. But at least we have a plan.

Pre-write checks

We can now list the five **Pre-write Checks**:

- **Who** are the expected readers?
- **What** do they want to see? (Basis of Decision)
- **Why** is my document special? (Unique Selling Point)
- **How** will my document be structured?
- **Where** will my content fit into the chosen structure?

Once you can answer these, the rest is easy. Most proposal-writers seem to crank up their word processor two minutes after they have read the ITT (or some of the ITT).

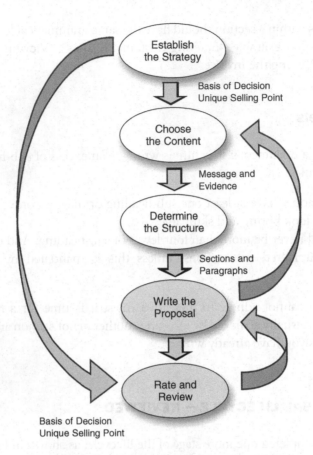

Establish
the Strategy

Basis of Decision
Unique Selling Point

Choose
the Content

Message and
Evidence

Determine
the Structure

Sections and
Paragraphs

Write the
Proposal

Rate and
Review

Basis of Decision
Unique Selling Point

FIGURE 4.3. Proposal lifecycle, stage 3 complete

They set down everything that comes into their heads, in any order, and expect the resulting jumble to win the business. But if you can answer the Pre-write Checks then you will have a clear idea of what you want to say and how it is to be structured, so your proposal will be focussed and persuasive. And that is writing that wins.

CHAPTER 5

Tightening up the text

SPIN THE WORDS

I now want to turn to the ways in which we can use the power of language to persuade our readers of the truth and beauty of our message. I quoted this example earlier:

> At present Management Information data is stored in a wide variety of different databases and the maintenance processes to collect and maintain the data are duplicated and inefficient. Several databases use outdated technology and consequently the applications to derive the data are also outdated. The aim of the new MIS Strategy is to unify together all these outdated databases into one central database. This will use up-to-date RDBMS technology which is maintained using a single data maintenance application.

What is wrong with this? The facts are all there; there's nothing wrong with the grammar, spelling or punctuation; and I'm sure we all understand what is being proposed. But it's so dull that before you get halfway through your eyes seem to have wandered off in search of something more interesting. It could be that the solution being presented is so novel and revolutionary that it demolishes any competing approaches, but that is not how it comes across.

Look at the example again. The writer has devised a strategy that needs to be explained and 'sold' to the reader. But the words chosen are repetitive, neutral and uninspiring. To be more convincing, the writer must do two things: make the existing situation look old-fashioned and slightly out of control, and make the proposed replacement seem full of potential. So, with these aims in mind, I'll give the text an overhaul. The result is as follows:

> MegaCorp management need timely and accurate information. Today, the statistics that support critical decisions are stored in many disparate files, so the procedures to keep them up to date are duplicated and inefficient. Our systems use obsolete equipment, and

the programs that calculate the figures that control the future of our company are outdated. The aims of the **New MIS Strategy** are to unify all the legacy files within one central database, to maintain the information accurately, and to exploit the modern retrieval and search technologies already widely used in our industry.

Table 5.1 shows what I have done, phrase by phrase:

All these changes were simple, but required a premeditated analysis of the emotions I wished to evoke in the reader. The remainder of this chapter describes more about how we can use the power and flexibility of language to spin our words – communicating the facts and keeping the reader's interest while building an emotional commitment to our solution.

The power of language

In *The Eve of St Agnes*, Keats writes:

> Full on this casement shone the wintry moon,
> And threw warm gules on Madeline's fair breast.

As Robert Thouless points out in his book *Straight and Crooked Thinking*, this verse is loaded with emotive language that contributes to the image the poet is trying to suggest. Words like *casement*, *Madeline* and *breast* are romantic and evocative choices. *Gules* is the heraldic name for red, but Keats has used his poetic licence to turn it into a noun that evokes both a time and an image: a sleeping maiden dappled with red-tinged moonlight. Without destroying the facts, we can neutralise the evocative words and 'translate' this verse as:

> Moonlight came through the window, making red streaks all over Maddy's chest.

Or, since Keats was once a chemist, he might have employed scientific language, like this:

> Lunar luminescence refracted by pigmented glass at 3000 Angstroms caused striation over the subject's mammae.

As this example shows, one of the major strengths of English is that it offers such a wide choice of words, each evoking a slightly different idea or image – for example, how does a *road* differ from a *street*, an *avenue*, a *lane* or a *boulevard*? You can alter the tone by changing the formality of the words (*appropriate*, *steal* or *pinch*), by using poetic language (*sweet* or *dulcet*), by utilising old-fashioned words (*raiment* or *clothing*), by deploying euphemistic terms (*passing away* or *dying*) or by being facetious (*imbibe at a hostelry* or *drink at a pub*).

Then you can influence the emotions by replacing neutral terms by those with more of a 'spin'. The classic example is that of a newspaper description of a *guerrilla* (a neutral term) as either a *terrorist* or a *freedom-fighter* – it all depends on the spin that

TABLE 5.1. How the example paragraph was spun

Old wording	New wording	Why?
(None)	MegaCorp management need timely and accurate information.	• Summarises the text that follows and draws it into a consistent argument. This is the attention-grabbing sentence at the top of the "paragraph triangle" described earlier.
At present	Today	• More immediate.
Management Information data	Statistics that support critical decisions	• Brings home the importance of the information in question; this is about *our* company and *my* decisions.
Wide variety of different databases	Many disparate files	• "Files" sound old-fashioned. • Stresses the incompatibility between the different data sources.
Maintenance processes	Procedures to keep these up to date	• Sows a doubt as to whether the existing information is reliable.
Duplicated and inefficient	(same)	• Good keywords to inspire management action.
Several databases	Our systems	• Removes an implication there is a problem in some areas but not others.
Outdated technology	Obsolete equipment	• Today we have "equipment", tomorrow we will have "technology".
Applications to derive the data	Programs that calculate the figures that control the future of our company	• "Data" is too abstract; we need to stress that this information is vital to the business. • "Programs" that "calculate figures" seem more old-fashioned than "applications" that "derive data".
The aim of the new MIS Strategy is	The aims of the **New MIS Strategy** are . . .	• Allows more structure to the list of benefits; further paragraphs could be structured around these three aims. • Capitalising "New" and the use of bold text give the package we are selling more of an identity.
Outdated databases	Legacy files	• Invokes an image of quill pens and ledgers.

TABLE 5.1 (cont.)		
Old wording	New wording	Why?
Use up-to-date RDBMS technology	Exploit the modern retrieval and search technologies already widely used in our industry.	• Removes jargon. • Hints at improved powers of decision-making and implies that competitors are already using these.
Maintained using a single data maintenance application	Maintain the information accurately	• A stronger hint that the current figures are inaccurate. • Lists the benefits in the order in which they will be obtained.

the reporter is giving to the story. A *businessman* might be presented as a *tycoon* or an *entrepreneur*; an *adventurous* idea might be considered as *bold* or *foolhardy*; an *unusual* solution can be described as *bizarre* or *mould-shattering*; or you might *criticise* a suggestion, *attack* it or *applaud* it. Consider this sentence:

> The report deals with the effects of the recent migration of the Payments System to the Oracle RDBMS.

Notice how the tone changes if the word "effects" is replaced by *impacts, achievements, consequences, aftermath, results, outcome, upshot, ramifications* or *repercussions*.

So there are all sorts of different ways to express an idea. You must pick the one that is most applicable to your readers, the formality of your document and the emotions you wish to evoke. To make our writing win, we have to harness the full power of the language. That means thinking not just about what we are trying to say but also about how to make those facts appeal to the specific readers who will be evaluating our message. There is no universal rule for this, although there are some useful tips in the remainder of this chapter. But the best method is to read and re-read what you have written, asking yourself if there are any more powerful words that will address the BOD, stress your USP and evoke the desired emotions in the reader.

AVOID ABSTRACTION

You want to bring your proposal alive, grabbing the reader so that they feel that they, and their needs, are being specifically addressed. So you need to avoid language that is distant, abstract and dry, like this:

> Objective consideration of contemporary phenomena compels the conclusion that success or failure in competitive activities exhibits no tendency to be commensurate

with innate capacity, but that a considerable element of the unpredictable must inevitably be taken into account.

That was George Orwell's satirical revision of a verse from Ecclesiastes, the original being:

> I returned, and saw under the sun, that the race is not to the swift, nor the battle to the strong, neither yet bread to the wise, nor yet riches to men of understanding, nor yet favour to men of skill; but time and chance happeneth to them all.

Orwell removed the races and battles, replacing them with abstractions and so producing something curiously similar to a corporate press release. The point is the same, but the translators of the King James Bible expressed it more vividly by taking specific examples and using simple language. Notice how the polysyllabic words in Orwell's version blur the message, even though we would expect the longer terms to be more precise.

You are writing a sales proposal, not undertaking an academic exercise or drafting a legal document, so put that tone out of your mind. Replace it with a confident, lively note, using simple language and examples that will animate your arguments. Profits may be up 36%, but if you start your analysis by saying so, it will immediately fall flat. Instead, bring the text alive by focussing on something everyone can relate to, for example:

> Last year we ordered 20,000 more stirrers for use at our coffee points. The reason: 120 new staff joined our communications division . . .

I'd read more of that.

In order to remove the abstracted tone that makes proposals so dull, we need to delve fairly deeply into the grammatical constructions that cause that effect. Once we know what to avoid, our arguments will automatically become more vigorous and more convincing. So the remainder of this section analyses the causes of fuzzy, flat writing, such as:

- Passive sentences
- Abstract subjects
- Weak words
- Dangling modifiers
- False subjects
- Negative constructions
- Vague figures

Passive sentences

Consider this sentence:

> Hackitout Software provided the most impressive proposal.

Here, the subject is "Hackitout Software", the verb is "provided" and the object of that verb is the "proposal". It is an **active** sentence, where the subject does something. Let's turn it round the other way:

> The most impressive proposal was provided by Hackitout Software.

The sentence is now **passive** – the subject and object have swapped around, so the subject, which is now the "proposal", has something done *to* it.

In general, passive sentences sound weak-willed, indecisive or even evasive. The active version is more natural and sounds self-confident. In particular, you should not use passive sentences in order to avoid first-person pronouns (*I, we, me* and *us*), as here:

> It is recommended that a further study be undertaken to explore some of these issues.

Instead, be self-confident with an active sentence:

> We recommend that we undertake a further study to explore some of these issues.

In this next example, nobody seems to want to take responsibility for the actions mentioned:

> Cost data will be collected and maintained to provide a detailed history of the hours spent in undertaking each stage of the work. This information will be entered into the project logging and tracking system.

Why not grasp the nettle? If we are going to undertake this work then let's be proud of it:

> We will collect and maintain cost data to provide a detailed history of the hours spent in undertaking each stage of the work. We will enter this information into the project logging and tracking system.

We are trying to persuade, not wriggle out, so be self-confident, honest and positive by using simple, active sentences with *we* as the subject.

The only valid uses of a passive sentence are when you have made a conscious decision not to mention the subject, when you don't know who it is, or when the object is more important than the subject. Examples of each such use are as follows:

> The PC blew up after a cup of coffee had been spilled over it.

> Unfortunately, the only backup file had been deleted three days earlier.

> The CORBA-standard Object Broker has been tested on the development system.

The Microsoft Word grammar checker can detect passive sentences. If you are happy with the ones it flags up then that's fine. But at least have a look at turning them into the active form. This can usually be achieved by swapping the positions of the subject and object, and then sorting out the resulting grammar.

Abstract subjects

Have a look at the subjects of these sentences:

> A new approach to the application of reusable core modules has made BigCo the number one supplier of mobile telephone billing systems.

> A request for all employees to submit their timesheets by the end of March is being made by the Finance Department so that customer accounts can be reconciled before the start of the financial year.

The subjects – *approach* and *request* – are vague, abstract nouns compared with the more concrete words elsewhere in the sentences. It is better to employ these more solid terms as the subjects – referencing people, organisations and places rather than abstractions like *situation, aspect, facility, issue, element, factor, matter* and *concept*. It is difficult for an abstract subject to undertake an action, so using one tends to lead to a passive sentence, as we can see from the second example. If we replace the abstractions with specific subjects, we can make the sentences more assertive, like this:

> BigCo is the number one supplier of mobile telephone billing systems because of its new approach to the application of reusable core modules.

> The Finance Department requests all employees to submit their timesheets by the end of March so that they can reconcile the customer accounts before the start of the financial year.

Weak words

Weedy, evasive writing utilises too many simple, weak verbs such as *is, are, was, were, can, could, has, had, have, do, did, done, make, get, use* and *come*. Obviously, these and other basic verbs are essential to any piece of prose, but if a stronger or more precise alternative is available then you should employ that instead.

Avoidance of strong verbs is a typical feature of the wishy-washy prose employed by petty officials, who think that plain speaking somehow weakens their authority, as here:

> The steering committee made a resolution that an investigation be carried out by a performance analyst into the feasibility of the provision of extra Oracle nodes.

The verbs are weak ones: "made", "be carried out". The words that *should* be verbs are all in there, but as nouns: "resolution", "investigation", "provision". Turning these into strong verbs gives:

> The steering committee resolved that a performance analyst should investigate whether it is feasible to provide extra Oracle nodes.

Here are some more examples of weak, evasive sentences with a more positive version of each:

1. We have made great improvements to our project management methodologies.

   We have greatly improved our project management methodologies.

2. The team's role is to perform problem location and resolution.

   The team locates problems and resolves them.

3. We would like to do more investigation before giving a full evaluation of the hardware options.

   We would like to investigate further before fully evaluating the hardware options.

4. Before the software implementation we will make technical and commercial evaluations of the benefits of the project, in conjunction with your own staff.

   Before implementing the software, we will evaluate the technical and commercial benefits of the project, in conjunction with your own staff.

5. Project Managers have to have an understanding of the skills of their team.

   Project Managers must understand the skills of their team.

It is not just verbs that can be weak. Adjectives such as *good, nice, bad, big, fast, easy, hard, slow* and *small* are also over-used. If you describe your disk capacity as "big" then the reader may just treat this as an empty boast – big compared with what? You need a more precise word or, in this case, to define the exact capacity.

Dangling modifiers

Examine the following example:

> After reviewing the cost/benefit analysis, the IT Manager decided to go ahead with the implementation.

There's nothing wrong with that. The first clause is a reference to the subject of the sentence, "the IT Manager", who was clearly the person who undertook the "reviewing". But now consider this sentence:

> After reviewing the cost/benefit analysis, a decision was made to go ahead with the implementation.

Again, we would expect the first clause to refer to the subject of the second. But a "decision" can't review anything – some unmentioned people did that. In grammar, this fault is termed a **dangling modifier** because you are left waiting for the subject to become apparent. Here's another example:

While undertaking the estimates, many risks were discovered.

Why did the writers of these sentences not wish to say who did the "reviewing" or the "undertaking"? As we saw when discussing passive sentences, it is all to do with shirking responsibility. When correcting the sentences, I will have to state who carried out these actions:

> After reviewing the cost/benefit analysis, <u>the IT Manager</u> decided to go ahead with the implementation.

> While undertaking the estimates, <u>we</u> discovered many risks.

I might have said "the meeting", or "the committee" or "the reviewer". The point is that I must be specific about who undertook the actions; otherwise the introductory phrase is left dangling. This is grammatically incorrect, evasive and over-abstracted.

False subjects

If a sentence starts "We decided that it ...", you would expect the pronoun "it" to represent something previously discussed. But now look at this:

> We decided that it was important for each system function to be fully documented.

Here, the "it" is a pronoun that should refer to something, but doesn't. It is what grammarians call a **false subject**. We want to avoid this sort of abstraction by homing in on the real subject, so by removing the padding we get:

> We decided that each system function should be fully documented.

This is still passive, so best of all is:

> We decided to document each system function in full.

Here are some more examples of false subjects – the corrected versions show how a sentence becomes clearer once they are removed:

1. In designing a redundant backup system, it is possible to meet the 99% uptime requirement despite the fact that there will always be a single point of failure.

   A redundant backup system will meet the 99% uptime requirement, although there will always be a single point of failure.

2. It will be possible, as the study proceeds, to identify performance degradations resulting from failure to upgrade the central systems.

   As the study proceeds, we will identify performance degradations resulting from failure to upgrade the central systems.

3. It will be the responsibility of the team leader to ensure a timely mix of skills from his staff pool.

   The team leader will ensure a timely mix of skills from his or her staff pool.

4. There are five factors that influenced our decision to repartition the system.

   Five factors influenced our decision to repartition the system.

Negative constructions

Don't use negation as a means of being evasive, as the writer does here:

The accounting procedures are not the most effective we have seen. The input processes are not universally liked, nor is the interface to the General Ledger using a very efficient protocol.

Instead, make definite, positive assertions, phrasing the sentence as follows:

The accounting procedures are ineffective, the input processes are disliked, and the interface to the General Ledger is inefficient.

Being positive also gives a better spin. Contrast these two sentences:

I hope you will not be disappointed with the test results but unfortunately the system will not be available for on-site acceptance until 3 September.

I'm sure you will be pleased by the test results and I'm glad to tell you that the system will be delivered for on-site acceptance on 3 September.

Vague figures

If you can provide explicit numbers, examples or precedents, your text will look much more convincing. A sprinkling of real facts might compensate for vaguer assertions elsewhere, so long as you make sure that they are relevant to the point being made, not just dragged in to bolster the credibility. Compare these two sentences:

Our new Babbage workstations have a large screen, several display modes and a small footprint.

Our new Babbage workstations have a large 15″ TFT screen, 12 display modes from 640 × 480 to 1024 × 768 pixels, and a small 6″ × 2″ desk footprint.

Numbers need to be quoted in context. If we say, "The processor is working at 95% capacity", it is unclear whether this is good news or bad. Better is, "The processor is working at 95% capacity against the target of 75%". Don't add an unnecessary level of precision, as in "Free storage capacity was down to 2,323,742 bytes on Tuesday at 2.31 p.m., so we urgently need a third 10.52 GB disk drive". And don't just present a mass of data, leaving the reader to fish out the conclusions. Interpret the information by turning it into charts or by highlighting the relevant figures. Finally, make sure that the figures are right, that totals are correct and that any values quoted are consistent throughout the document. Numbers can add credibility and persuasiveness to proposals, but if they are wrong or puzzling then the effect is reversed.

Summary

We have now seen what happens when writers try to avoid over-commitment. Passive sentences, dangling modifiers and the other abstraction techniques come into play, and the result is vague, dry and unconvincing. Nobody will be fired for producing such prose, but not much new work is going to be won, either. It is not writing that convinces or persuades.

However, sometimes we do need to be a little evasive: some aspects of our solution are untested, key staff may not be available, there might not be time to implement everything we describe. But you can be slippery without sacrificing clarity. Instead of adopting an abstracted, noncommittal tone, wriggle out with words like *may*, *might*, *could*, *should*, *hope to*, *aims to*, *perhaps*, *indicate*, *suggest*, *appear*, *seems*, *usually*, *likely*, *possibly*, *probably*, *normally*, *in most instances* and *generally*.

But don't wriggle unless you *have* to wriggle. You want to make your proposal sound as authoritative as possible. Look at this example:

> At this stage, the working assumption is that the Hackitout product may provide the standard CORBA implementation for use throughout most, or possibly all, of the new MegaCorp enterprise-wide architecture.

Here, I happen to know that a decision to use the Hackitout product had already been made – indeed by the writer himself. But, as is all too apparent, he is scared that his decision may one day prove to be incorrect. Such fudge gives a very poor impression in a proposal – is this a recommendation or not? If there are some doubts or caveats then you should state exactly what they are, but the overall tone should always reassure the reader that you understand the problem and are confident about your solution. The text should just have read:

> The Hackitout product will provide the standard CORBA implementation throughout the new MegaCorp enterprise-wide architecture. However . . .

Any doubts can then be listed without destroying the note of authority that this initial statement establishes.

STICK TO THE POINT

Why does good prose suddenly ramble off into irrelevant, boring, long-winded waffle? I can think of three main reasons: personal hobbyhorses, the attraction of irrelevant detail and simple ignorance of the topic.

Hobbyhorses

Everyone has a personal swarm of bees in their bonnet and these have a habit of buzzing in, regardless of their applicability to the issue in hand. For example, one ex-colleague of mine was obsessed with reconstructing her organisation's database architecture in a particular way, so her reports, regardless of the subject, somehow always led to this same conclusion. And I am sure we are all familiar with the evangelist who has read up on the latest technology and bores you with this, rather than finding a solution for the current problem. Curiously, it is easy to recognise hobbyhorses in other people's prose, whilst your own remains mysteriously free of prejudice and perfectly focussed. It's not true – so make sure that your key arguments are reviewed and agreed before too much text has been written.

True but irrelevant

That you know something is not a reason to include it. There is an understandable yearning to pepper pages of speculation with a few hard facts, adding to your credibility by throwing in one of your few crumbs of actual knowledge. Fine, if it is relevant, but too often I read pages of less-than-fascinating data that does not add anything to any argument that the writer is trying to make. Undoubtedly, it is all true, but someone should have asked how this text is meeting my Basis of Decision or illuminating a Unique Selling Point. As I stated earlier, there is always a reluctance to axe words once they have been written. So try not to write them.

Don't understand

We have all been in situations where a paragraph or two is needed on something we don't understand. Maybe the customer asked us to cover a particular topic, or maybe we *should* discuss that topic but have nothing sensible to say about it. All I can advise is to keep it short. You will only ever fool the equally ignorant or the speed-reader, so don't think that if you provide a long section crammed with buzz-phrases then you'll not be rumbled. You cannot disguise ignorance with fuzzy writing. When a clear, sharp and incisive document suddenly cuts to information-free babble, I know I have reached the part the writer is worried about. I then know what to question and contest at a review.

If you have been asked for your opinion, don't fudge around all the possibilities because you haven't reached a decision. Either pick one, or say why you haven't; then you can write your report in confident, plain-speaking prose.

USE PLAIN LANGUAGE

> I shall say that, as far as we can see, looking at it by and large, taking one thing with another, in terms of the average of departments, then in the last analysis it is probably true to say that, at the end of the day, you would find in general terms that, not to put too fine a point on it, there really was not much in it one way or the other.
>
> (From *Yes Minister*, written by Antony Jay and Jonathan Lynn)

There are several ways to turn plain, simple writing into gobbledegook. Among these are technobabble, buzzwords, legalese, Pentagonese, clichés and windy phrases – each of which is discussed in this section.

Technobabble

A few weeks ago, my train came to an unexpected halt at a remote country station. Over the loudspeaker came a muffled, distorted request for "all customers to immediately detrain the unit". Nobody moved. A few moments later, an angry guard stormed through the train asking us why we hadn't got off. What that guard (or the passenger service manager as he is now termed) had done was to employ the jargon that he used when talking to his colleagues, assuming that it would be understood by the passengers.

The world of IT is particularly prone to jargon. And a lot of it is essential. We refer to things like modems, Bluetooth and 2.8 GHz CPUs because that is what these things are called. There are no simpler terms. My mother doesn't understand them, but I'm sure that most readers of this book will do so. However, jargon can sometimes be used to disguise fuzzy thinking, creating a sort of technical fog that no one dares to penetrate. This just alienates the reader, as does any implication that a concept can only really be understood by fellow-geeks. Technobabble is used as a kind of secret sign that implies "no entry to the ignorant". This completely misses the point of what we are trying to do – communicate our ideas.

TABLE 5.2. Buzzwords		
Best practices	Initiative	Parameter
Blue-sky thinking	Interactive	Process-based
Challenge	Interface	Pushing the envelope
Constructive	Joined-up	Scenario
Dimension	Learning curve	Stakeholder
Facility	Ongoing	Synergy
Holistic	On-stream	Track record
In-depth	Paradigm shift	Workshop

Nor should you invent new words. Someone once observed there isn't an English noun that can't be verbed. So in the following examples we can understand what the writers are *trying* to say:

We must commit to obsolete the flat-file data formats as soon as possible.

We have no one working on this capability at present, so we do not know when it will event.

However, the fact is that *obsolete* and *event* are not verbs – and there are plenty of more literate ways of making these points. If you keep verbs as verbs and nouns as nouns, you don't end up with some contorted phrase like "to deliver the implementation of" when you can just say "to implement".

And because you can't invent new words, you must eliminate jargon that is not valid English, like that used in these examples:

You can c&p the instrument code directly into the yield-calc box and go it from there.

BigCo has formed a global alliance with MegaCorp with the objective of delivering joint business solutions to customers in the financial services vertical.

There are times when a good solid chunk of technobabble is essential in a proposal. It is aimed at a specific audience, and shows that we understand the technology being proposed. But such sections are the exception. The bulk of the text must use simple, direct, short terms that every reader will understand.

Buzzwords

Another way to cloud your meaning is to dress up commonplace concepts by making them look fresh and state-of-the-art. Empty words like the ones in Table 5.2 show

TABLE 5.3. Legalese expressions	
Acquaint	In receipt of ("I am in receipt of your
Aforementioned, aforesaid	letter of 13 May.")
Apprise	In respect of
As per	Monies
Cognisant	Perchance
Deemed	Persons
Duly ("If you duly apply on the enclosed form.")	Said ("As defined in the said
Forthwith	document.")
Furnish ("We have not yet been furnished with the	Therein, thereto, thereat, thereby,
detailed specification.")	thereof
Herein, herewith, hereto, hereat, heretofore, hereby	To hand ("I have your letter to hand.")
Hitherto	Under separate cover
Inasmuch as	Whensoever
In lieu of	Wherein, whereby, whereof

that you are just stringing phrases together, not thinking about the customer's problem and how best to communicate your solution.

If the entire proposal is filled with this kind of guff (and I have seen many that are), it looks lazy and unconvincing. Use plain words and your readers will trust you more.

Legalese

Musty old words

Examine this example:

> This document specifies the requirements to create a relational database containing details of global base metal smelting and production capacities. It is expected that a formal Functional Specification can be produced from the information furnished herein.

Why does it end with a phrase from a Victorian will? It just makes the writer seem pompous and old-fashioned. We are not drafting Acts of Parliament, so avoid words and phrases like those shown in Table 5.3.

There is also no need to repeat numbers, as in "This quotation is valid for ninety [90] days". It is a convention from times when figures in handwritten contracts were open to a little manual adjustment. Now it is obsolete.

Use of 'we'

In a proposal, it is best to use *we* and *you* to refer to your company and your customer respectively. I suspect some writers think this makes the document too informal and hence less legally binding. But believe me, if a proposal is issued by your company and says *we* will do something, the judge isn't going to think it means anyone else. An initial use of company names is usually unavoidable, but try to get into the touchy-feely stuff as soon as you can, like this:

> Hackitout Software is pleased to present our proposal to MegaCorp Ltd for . . . We have developed this solution in conjunction with your own staff . . .

Now look at this example:

> Hackitout Software will write a User Manual, and agree this with nominated representatives of MegaCorp Limited. Hackitout Software will also train MegaCorp staff in the use of the system during three hands-on sessions supervised by Hackitout trainers and attended by officially designated MegaCorp operators.

An entire company is too big and too distributed to write a manual. More importantly, its sales staff have probably spent months in building up a personal relationship with this customer and would like to keep that cuddly image, rather than that of a big, impersonal organisation. And what's all this "nominated representatives" and "officially designated" stuff? More legalese. If we are training the customer's staff then it is not necessary to say that the trainers are our employees and the trainees are the customer's. So, in plain English:

> We will write a User Manual and agree it with you. We will also train your operators in the use of the system during three supervised 'hands-on' sessions.

The one place where *we* and *I* should be removed is in phrases like "We believe . . . ", "In our view . . . ", "Our proposal is to . . . ", "I am of the opinion that . . . " and "My own preference is to . . . " Just cut these out and leave the text that follows as a statement. The reader knows the proposal represents your opinion, your approach or whatever, and it weakens the argument if you 'protect' it with such an introductory remark. See how much more authoritative the following text becomes once we remove the caveats:

> We believe that the Toronto databases are slow, inaccurate and difficult to change. Our suggested long-term strategy is to incorporate the information from these into the new MBAS system currently being created in London. In our view, each record will need to be thoroughly checked and cleansed before this process can start.

> The Toronto databases are slow, inaccurate and difficult to change. The best long-term strategy is to incorporate the information from these into the new MBAS system being created in London. Each record will need to be thoroughly checked and cleansed before this process can start.

Shall and will

Some people believe that *shall* implies a legal commitment but *will* just implies a desire or hope. This is false. The use of *will* in the following example is as much of a commitment as saying *shall* or *must*:

> The user will receive a response within 1 second.

The old rule was to use I <u>shall</u>, you <u>will</u>, he she or it <u>will</u>, we <u>shall</u>, you <u>will</u>, they <u>will</u>, then to reverse the *wills* and *shalls* when writing promises, commitments, or commands. Unfortunately, few people still apply this rule and you'd have a hard time getting it past a judge. If you want to commit to something then use *must* or *shall*. If you don't want to commit then it's time for some weasel words like *may* or *might* – but *will* is not weaselly enough.

Other legal minefields

There are a couple of stock phrases with legal interpretations that you might not wish to take on.

- **Best endeavours**. Some people use this term as meaning, "We'll do it if we can be bothered but don't hold your breath". In fact, it means that we will make every conceivable attempt to do something, including diverting our entire workforce onto the problem. Something like "all reasonable efforts" at least gives grounds for dispute.
- **Time is of the essence**. This acknowledges that we understand that the stated delivery dates must and will be met. If necessary, we will allocate all the company's resources to meet these dates, and will expect to be sued for all consequential losses if we fail. *Never* use this phrase.

Pentagonese

Pentagonese is the ability to call a spade a "terrain redeployment asset". In the next example, the writer is too grand to be talking about an unimportant subject like backups, so dresses it up with a bit of abstraction:

> Continual ongoing expansion of the General Ledger Archive Log has rendered the residual RD1000 resources insufficient for periodic and aperiodic recovery allocations.

Er . . . so does that mean we're out of disk space?

> The aim of the workshop session will be to determine time horizons for the implementation of the solution.

... and we're all going to sit round and make a guess.

> Because of the fluctuational predisposition of your position's productive capacity as juxtaposed to government standards, it would be momentarily injudicious to advocate an increment.

... so no payrise for you.

What do you think about the writers of those last three examples? Cleverer than you? Surely not. So why not use simple words? For instance, replace *alleviate, apprise, endeavour, manifest, necessitate* and *avail oneself of* in cases where you mean *ease, inform, try, show, need* and *use*. You won't look stupid, and your readers will thank you for it.

However, there's nothing wrong with using a longer word if it adds precision to what you wish to say. By limiting your language, you also limit your freedom to express the subtleties of your argument. It is weak to say that your solution is "good" – it may be *ingenious* or *practical* or *appropriate*. *Design* is a simpler word than *configuration*, but "the configuration of the workstation" is very different from "the design of the workstation". However, if we were to say, "Most manufacturers of keyboard cables utilise a spiral configuration" then pomposity is beginning to rear its ugly head. And talking of clichés . . .

Clichés and metaphors

> Manpower ceilings are a very blunt macro-instrument and will be unduly restrictive if not based on the result of management reviews and other 'micro' activities. Ceilings are biting, but this is what they were meant to do.

> We have reached the limits of propeller-driven technology with the relational database, but we have a huge investment in our current systems and need to keep this craft aloft while we prepare for in-flight refuelling and eventual rescue. However, the handwriting is already chiselled indelibly on the wall.

Most clichés are worn-out metaphors. I'm sure that in the seventeenth century the first person to use the phrase *avoid like the plague* was considered quite a wit. Now, plague is an unlikely occurrence in the life of the modern office worker, but *avoid like legionnaire's disease* just seems in bad taste. So it is better to use a cliché than some phrase of your own that sticks out like a warty nose.

Clichés and clichéd metaphors are too numerous to mention. It would be impossible to avoid them all – I'm sure you can find plenty in this book. But there is nothing wrong with a metaphor, however clichéd, so long as it is appropriate. I might say:

> Writers of business reports would find eliminating *all* clichés a hard row to hoe.

But in the context of a guide for white-collar professionals, the use of a metaphor from farm work doesn't quite work. In contrast, the other day I wrote something like:

> Despite the Herculean efforts of the data clerks, there is still a backlog of nearly 2000 transactions awaiting entry into the ledger system.

Obviously, the data clerks have never cleansed the Augean stables, but the use of "Herculean efforts" made my point more vivid and more 'imaginable' than some bland phrase like "hard work".

Mixing a metaphor is to overload a sentence with too many disparate images. In *The Moor's Last Sigh*, Salman Rushdie writes:

> The moon, a silver coin hung in the draperies of the enchanted night, let fall her glance, which gilded the rooftops with a joyful phosphorescence.

A beautiful description at first reading, but are coins normally found hanging in draperies? How can something silver be used for gilding? Can phosphorescence be joyful? Is Rushdie pulling our leg? Anyway, you don't have to use 'poetic' language to mix a metaphor, as shown by this example:

> The course allows students to cultivate the strategic foundations of an OO deployment, and illustrates some weapons and tactics.

That course also seems to cover architecture – or is it horticulture, or military planning?

It is the *thoughtless* use of metaphors and clichés that make that example, and those at the start of this section, so ridiculous, and why Sam Goldwyn once advised that you should "Avoid clichés like the plague". So don't just create clichéd text while writing on 'autopilot'. Instead, use metaphors sparingly and consciously, choosing those that are appropriate and which effectively enhance your argument.

Windy phrases

Here we are talking about the use of several words when one will do. Why say *postponed until later* – when else would it be? Why say *specific example* – all examples are. Why say *ultimate end* – where else would an end be? We don't "come to an agreement", "give consideration to" or "conduct an investigation": we *agree*, *consider* and *investigate*.

Here's an example of windy writing:

> The basic fundamental of the strategy is to connect together the existing screens to a database which is larger in size and, over a period of time, replace the old one with an exactly identical copy.

With use of an axe, it can be made much clearer:

> The basic fundamental basis of the strategy is to connect together the existing screens to a larger database which is larger in size and, over a period of time, replace the old one with an exactly identical copy.

Here are some more examples of windy sentences, with a more precise alternative for each.

1. A 20 MB allocation of server space is in the process of being set up for you.

   I am setting up a 20 MB allocation of server space for you.

2. For the benefit of new joiners, the manager described the structure of the department and the remit that had been given to it.

   For the benefit of new joiners, the manager described the structure of the department and its remit.

3. I would like to take the opportunity to apologise for the delay in replying to your request.

   I apologise for the delay in replying to your request.

4. This letter is to advise you that the version of the MATRIX software you are currently running will no longer be supported from the end of the month.

   From the end of the month, we will no longer be supporting the MATRIX software you are currently running.

5. In view of the fact that the problem still remains, we will cancel our January invoice.

   The problem remains, so we will cancel our January invoice.

6. The Alert Notification facility permits the operator to immediately become aware of emergency situations.

   The Alert Notification facility warns the operator of emergencies.

Avoid saying the same thing twice, as in this sentence:

> Finally, we believe in conclusion that this is an excellent solution.

And why have paragraphs or sentences starting "Note that" or "N.B.", or section headings of "Notes"? Are we not supposed to have taken any note of what has gone before? You may use "Notes" to append a number of points after a diagram or a quotation, but you cannot annotate your text with *more* of your text.

Tight, punchy writing contains no excess flab. If words can be removed from a sentence then you should have no qualms in striking them out.

Summary

Plain language is achieved by turning on a kind of mental filter. As you write each word or phrase, you should be asking yourself the following questions:

- Will my readers understand this?
- Is this clichéed or long-winded?
- Is there a shorter, simpler or more modern way of expressing this?
- Is there a more specific word or phrase that I could use for this?

Skilled and experienced writers apply this filter all time, and without thinking too much about it. It is a talent worth acquiring.

CONSIDER THE TONE

In verbal communication, the tone in which you say something can often be as important as what you say. For example, a simple "hello" can be given a wide variety of intonations: friendly, sneering, bored or surprised. Written communication, too, has a tone, which emerges both from the content and from how it is expressed. For example, I once reviewed a proposal that started by haranguing the customer for only allowing two weeks for its preparation. In itself, this was bad enough, but it gave the remainder of the document a begrudging and aggressive tone that was not helpful in winning the business.

We must address the reader with respect and politeness, avoiding arguments and phrases that are in any way jarring or inappropriate. Here are some rules that may seem obvious, but which I have seen broken more than once:

- Don't tell the customer about their own business. You should show that you understand the problems you are trying to solve, but there's no need to trace these back to the foundation of the company or their last balance sheet.
- Don't replace the customer's terminology with your own. I often see attempts to replace customer jargon with terms from a vendor's standard product. It just gives the impression that the vendor is trying to weld a problem to an inappropriate solution. If you repeat the very words the customer has chosen to express their problem then they will be more inclined to believe that you understand what is wanted.
- Don't overplay any personal relationships that you have built with the customer's staff. This may embarrass the employees concerned, and you may well find that the relationship is not as warm as you supposed. Above all, you must not imply that any such rapport has yielded insight or information that is not available to the competition.

- Don't state that the customer's requirements are stupid, unworkable, technically inept, outdated or unnecessary.
- Don't be too simplistic. Just because you are targeting management readers, you shouldn't assume that they can't understand a diagram without having every component described in words of one syllable, or that they need acronyms like "CPU" to be explained.
- Don't attack the competition – especially by name. You can point out the weaknesses of a particular approach and let the readers draw their own conclusion, but it is better to concentrate on the strengths of your own solution.
- Don't be excessively self-promoting. In particular, don't let the glowing description of your company's philosophy, reputation or products overwhelm the explanation of the specific solution.
- Don't criticise the customer, for example by highlighting past delays or inappropriate decisions. Those decision-makers may be evaluating your proposal.
- Don't be over-optimistic. If you say you are going to do something or supply something then you must say how, or at least *know* how, you intend to do it.
- Don't make empty claims. A line like "Our Mark II widget is the most reliable on the market" only makes sense if backed up by facts.
- Don't lie, exaggerate, misinform or falsify. A proposal is a legal document and, in any case, if your lies are exposed then your argument is lost.

A good test for the right tone is to mentally read the text 'aloud'. Ask yourself if you would like to be addressed like this. Imagine how your customer or manager would feel. If the tone seems patronising, aggressive, disrespectful or evasive then it is time for a rewrite.

REMOVE THE BOILERPLATE

Many companies have an extensive library of pre-packaged lumps of text that can be slotted into any proposal. Typical boilerplate elements include the company history, previous financial results, essays on current industry trends, summaries of previous projects, staff résumés, commercial terms and conditions, quality procedures and product fact-sheets. In addition to these, text can be recycled from previous proposals, allowing a completely new document to be quickly assembled from existing components.

That is the theory, anyway. The truth is that the proposals that result are usually dull, unconvincing and out of date. Long stretches of boilerplate are easily detected and skipped by readers, while the nuggets of specific information – like the solution or the price – are buried. The reader is left with the impression that you think every customer is the same, and that you can't be bothered to tailor a solution to their particular needs.

Firstly, you should be asking whether the boilerplate should be included at all – does it meet the BOD or bring out your USP? Secondly, can the information be shortened – do your readers really want full staff résumés and every detail of some previous projects? Finally, can the text be personalised – made relevant to the expected readers and their specific problems? For example, the following sentence is typical thoughtless boilerplate:

> We have extensive experience in communications systems and in the provisioning and monitoring of complex networks.

That may be true, but the words must be tuned to address the individual circumstances:

> Our experience with complex communications systems will allow BigCo to provision networks quickly and cost-effectively, and to monitor them so that problems like the recent outage in Coventry can be speedily resolved.

I'm not saying that boilerplate text and sections recycled from previous proposals have no place. Use of such material will speed up the production process and give a consistent message to different customers. But remember the rule: every word must be targeted to the BOD and USP *for this specific proposal*.

REMOVE THE CROSS-REFERENCES AND FOOTNOTES

Nobody ever looks up cross-references. Suppose that the referenced section itself referred you elsewhere – you might never finish. What a cross-reference usually indicates is, "This topic is discussed or expanded on elsewhere, so don't worry". Your chosen document structure should be presenting all the information in a logical order, so if you find you need a multitude of cross-references it is a sure sign that the structure is wrong.

Don't use *above* and *below* to refer to other sections of your document. A phrase like "as discussed in the above chapter" is grammatically incorrect and vague and may be forgotten if the text is restructured. For the same reason, care should be taken with referencing *preceding*, *next*, *earlier*, *later* or *previous* sections or diagrams.

Footnotes should also be used with caution. They can irritate the reader, who never knows whether it is worth being distracted from the main point only to find a reference to some document that they will never read, or some irrelevant metaphysical musings by the author. Either something belongs in the text or it doesn't.

TUNE EACH SENTENCE

Let us assume that we have now spun the text to make our points count, we have avoided an over-abstracted tone, we are sticking to the point, using plain language and have removed any unnecessary material. How else can we improve our proposal? It is time to examine each individual sentence to see that it is well constructed and of an appropriate length. After fine-tuning each sentence in this way, we can finally say that we have made our completed text as clear and effective as possible. In this section, I will examine the following:

- Sentence length
- Sentence focus
- Repetition of words
- Use of the little words

Sentence length

Some people seem to be under the impression that full stops are taxable, for example:

> Object modelling is currently going through a transition, with the mainstream object oriented modelling packages merging to form the Unified Modelling Language (UML), which aims to take the best from all the competing models, so enabling products to settle on one modelling method and unifying the different syntax and notations of OMT and UML, with most tool manufacturers incorporating the new notations in their products, although not all the changes have or will have been added within the timeframe of this project, which will be based on a combination of OMT and UML notation with the latter having preference.

Phew! Here are some tips for shortening your sentences:

- **Remove the waffle.** Cut out any words that are not contributing to your point.
- **Split it up.** Look for the natural break points, such as *and*, then start a new sentence there instead. If the split is too abrupt then you need a connector between the two sentences, such as *however* or *in addition*.
- **Remove the repetition.** Replace repeating terms with pronouns, such as *it*.
- **Use a list.** Place all the supporting points in a bulleted or numbered list.

You want to aim for a mixture of sentence lengths, averaging around fifteen to twenty words. For example, here is Richard Dawkins from his defence of evolutionary theory, *The Blind Watchmaker*. Notice how he begins with a short, attention-grabbing sentence, then, among some longer descriptive sentences, he inserts briefer, factual points that contain the real meat of his argument.

> It is raining DNA outside. On the bank of the Oxford canal at the bottom of my garden is a large willow tree, and it is pumping downy seeds into the air. There is no consistent air movement, and the seeds are drifting outwards in all directions from the tree. Up and down the canal, as far as my binoculars can reach, the water is white with floating cottony flecks, and we can be sure that they have carpeted the ground to much the same

radius in other directions too. The cotton wool is mostly made of cellulose, and it dwarfs the tiny capsule that contains the DNA, the genetic information. The DNA content must be a small proportion of the total, so why did I say it was raining DNA rather than cellulose? The answer is that it is the DNA that matters. The cellulose fluff, although bulky, is just a parachute to be discarded. The whole performance, cotton wool, catkins, tree and all is in aid of just one thing and one thing only, the spreading of DNA around the countryside. Not just any DNA, but DNA whose coded characters spell out specific instructions for building willow trees that will shed a new generation of downy seeds. Those fluffy specks are, literally, spreading instructions for making themselves. They are there because their ancestors succeeded in doing the same. It is raining instructions out there; it's raining programs; it's raining tree-growing, fluff-spreading algorithms. It couldn't be any plainer if it were raining floppy disks.

That has 255 words in fourteen sentences, an average of just over 18.

Although excessively long sentences are the more common offence, you can err too far in the other direction. Only Hemingway could get away with this (from *An Alpine Idyll*):

The sun came through the open window and shone through the beer bottles on the table. The bottles were half full. There was a little froth on the beer in the bottles, not much because it was very cold. It collared up when you poured it into the tall glasses. I looked out of the open window at the white road. The trees beside the road were dusty. Beyond was a green field and a stream. There were trees along the stream and a mill with a water wheel. Through the open side of the mill I saw a long log and a saw in its rising and falling. No one seemed to be tending it. There were four crows walking in the green field. One crow sat in the tree watching. Outside on the porch the cook got off his chair and passed into the hall that led back into the kitchen. Inside, the sunlight shone on the empty glasses on the table.

Again fourteen sentences, but just 164 words, an average of 11.7. Hemingway is able to use this pattern to evoke an impression of a slow, lazy day (can you see how?), but in formal writing such a style would seem jerky and unconnected.

Microsoft Word will 'green-line' any sentence longer than 60 words. Sometimes this can be ignored – where you have a long list, for example. But usually there are better ways to get your point across.

Sentence focus

A sentence should just contain a single logical thought, so you should try not to weld several different ideas together, as the writer does here:

We considered how best to present the conflicting views of the staff we interviewed, some of whom we travelled to New York to see, with the sales staff being the main dissenters from the central view that is described in Section 5.

Make sure that you don't lose sight of the subject or verb in the sentence, as happens in these examples:

> The CEO will be giving a presentation on Friday in the large meeting room to which all staff are invited on "The Year Ahead" at 2pm.

> The summary record in the log file from the statistics library on the second disk in the array was accessed only once last year.

In that last example, the verb comes so late in the sentence that the subject has been forgotten. You could make the description into an addition to the sentence, which would then become:

> One record was accessed only once last year – the summary in the log file from the statistics library on the second disk in the array.

Where a sentence seems long and confusing, try stripping out the 'decorative words' – the adjectives and adverbs – and make a simple sentence out of what is left. Then decide which additional words are essential to the meaning before re-introducing them.

Repetition of words

Wordsworth's *The Solitary Reaper* begins as follows:

> Behold her, single in the field,
> Yon solitary Highland Lass!
> Reaping and singing by herself;
> Stop here, or gently pass!
> Alone she cuts and binds the grain . . .

I think he makes it more than clear that there's no one else around. The poet uses four different terms – "single", "solitary", "by herself" and "alone" – to establish this fact. Over-use of the same word seems plodding and unimaginative, as shown by our old friend:

> At present Management Information data is stored in a wide variety of different databases and the maintenance processes to collect and maintain the data are duplicated and inefficient. Several databases use outdated technology and consequently the applications to derive the data are also outdated. The aim of the new MIS Strategy is to unify together all these outdated databases into one central database. This will use up-to-date RDBMS technology which is maintained using a single data maintenance application.

This clunks along because some words – *database*, *maintain*, *outdated*, *technology* and *application* – appear more than once in this short introductory paragraph. The effect is to make the ideas seem boring and uninspiring.

Unfortunately, Microsoft Word won't spot repetitions (why?), but it does provide a thesaurus. So if you revise a paragraph and find that the same word keeps banging you in the eye, try to find an alternative. For example, in a paper I recently wrote on database strategy, the word *data* naturally occurred many times. The thesaurus suggested the alternative of *information*, so by using this instead of *data*, where I could, I made the prose less stodgy.

As already discussed, there are times when a word or phrase *should* be repeated – to flow ideas within and between paragraphs. As a reminder, consider this argument:

> Many people use the spelling checker in Microsoft Word but ignore its excellent facilities to validate grammar. If writers enabled the grammar checker, and acted on its recommendations, the quality of our proposals would be greatly improved. And proposals are the life-blood of our business.

Here I have carried forward the words "grammar" and "proposals" so that the flow of the argument is maintained, but I used "validate" in the first sentence to avoid excessive repetition of more "check".

You should not sacrifice consistency just because it would mean repeating a phrase. In this example, the writer has tried to define a term, but failed to nail it down because the same wording is not used.

> **Deliverable Accepted** status is obtained once all customer signatures have been obtained. The project log is updated with the date on which the agreement was achieved and the Delivery Agreed situation is then notified to all relevant project and client staff.

Here is another example:

> **Intent**
> The overall strategic goal is to replace the New York databases with more modern systems hosted in London. This ambition drives all the lower-level objectives, which in turn shape the design and, ultimately, the implementation of the platforms.
>
> **Aims**
> Three specific high-level architectural goals shape the principles that underlie the design . . .

The use of different terms (*overall*, *lower-level*, *specific* and *high-level*), (*intent*, *goal*, *ambition*, *aims* and *objectives*), (*databases*, *systems* and *platforms*) obscures the intention, which is to make the second section into a lower-level breakdown of the "overall strategic goal". The writer has done a good job in avoiding repetition, but needs to define some terms and announce the structure.

Inconsistent terminology is a particular peril when documents are authored by more than one person. It often seems petty to ask someone to alter some detail of wording – for example to change "Delivery Agreed" to "Deliverable Accepted" – but it makes a big difference to the reader.

The little words

If a sentence doesn't look quite right, there are two ways to improve it. One is to adjust the phrasing by use of commas, which I describe in the chapter on punctuation. The other is to adjust the little words, often by repeating *the*, *a* or *an*, or inserting a clarifier such as *so that*, *then*, or *in order to*. These tiny adjustments can remove ambiguity and improve the flow, without a wholesale revision.

Here I am trying out some small adjustments to see if I can improve the clarity of my argument.

1. This document describes the current problems being experienced, and the stages by which it is proposed to solve these <u>using</u> architectural elements compatible with the overall strategy.
2. This document describes the current problems being experienced, and the stages by which it is proposed to solve these <u>by utilising</u> architectural elements compatible with the overall strategy.
3. This document describes the current problems being experienced, and the stages by which it is proposed to solve these <u>by the utilisation of</u> architectural elements compatible with the overall strategy.
4. This document describes the current problems being experienced, and the stages by which it is proposed to solve these <u>through the utilisation of</u> architectural elements compatible with the overall strategy.

Sometimes you need additional words to keep the sense, as here:

The presentation by our rivals was notable for its disagreement and disdain for the client's requirements.

In that example, the "for" is trying to serve "disagreement" and "disdain", but "disagreement for" is not a valid construction. Better is:

The presentation by our rivals was notable for its disagreement with, and disdain for, the client's requirements.

Often, two or more parts of a sentence need to be kept in parallel, using the same grammatical format, but then lose the little words that make this happen. The following two examples demonstrate this:

From this menu, you can choose which reports to run now and the ones to queue for a later batch run.

The last three years have given me extensive practice in working under pressure and to meet tight deadlines.

In the first example, "the ones" should be replaced by another "which". In the second, "to meet" should be replaced with "in meeting". In both cases, the revised text gives the two options a parallel construction.

Now consider this example:

> The study will undertake a number of 'what if' assessments to choose between the
> different technical, environmental and functional options.

Using "in order to" rather than "to" links the second part of the sentence more clearly
to the verb "undertake". I use this trick often.

The little words can also be used to change the spin. Consider this example:

> At present, the architectural design concentrates on Strand 1, although the overall
> strategy will permit the other strands to be incorporated later.

"Although" is not a good choice here; it creates an apologetic tone for the remainder of
the sentence and leaves the impression that the "other strands" don't matter much.
Better is:

> At present, the architectural design concentrates on Strand 1; the other strands will be
> incorporated later, within the overall strategy.

If you can't make a sentence read convincingly by playing around with either the
commas or the little words, then it is time for a total rethink.

CHAPTER 6

Obeying the grammar rules

DOES GRAMMAR MATTER?

My knuckles still feel a ghostly pain whenever I break the writing rules beaten into me at what, significantly, was called a "grammar" school. In those days, grammar *did* matter. Perfect prose carried more weight because it instantly indicated that the writer was "one of us" – someone who understood the rules and hence was to be trusted. Some remnants of that attitude persist. There are still some people who, should they encounter a split infinitive or some other manifestation of incorrect grammar, immediately bin the offending text. Others may decide that the writer is only semi-literate, but will soldier on, being convinced more by the arguments than by a failure to obey some arcane rule. But, whoever the readers may be, we don't want to interrupt their understanding of our world-beating technical solution with something that makes them pause, wondering if they have read it right.

Unfortunately, when you ask most people if they know any rules of grammar, they will come up with something trivial. In fact, most of the rules you think you know are wrong: you *can* split an infinitive, end a sentence with a preposition or start one with a conjunction. In this chapter, I will show why. But those rules don't matter; the situations where they arise are rare, and the resulting English is not so bad anyway. What concern me more are the glaring, fundamental errors that immediately devalue a piece of text and destroy my concentration on the arguments being made. These are the grammatical rules I see broken every day, and which I want to clarify in this chapter.

Rules you can break

Don't start a sentence with a conjunction

Theory says that conjunctions such as *and, but, or, because* and *so* are reserved for linking thoughts within a single sentence. But this is rubbish. Such words are very effective in joining successive sentences together, particularly when you are making an argument. And they gain in power by being used as the first word. Jane Austen often starts sentences with *but*, as here:

> She had two sisters to be benefited by her elevation; and such of their acquaintance as thought Miss Ward and Miss Frances quite as handsome as Miss Maria, did not scruple to predict their marrying with almost equal advantage. But there are certainly not so many men of large fortune in the world, as there are pretty women to deserve them.

Don't split infinitives

An infinitive is the form of a verb starting with *to* – for example, *to analyse, to program, to crash*. A **split infinitive** occurs when you place a word between the *to* and the verb. In a sentence like "We easily brought the project in on time", there is no split infinitive because "we brought" is not the infinitive form. But "It is rare to easily bring a project in on time" *would* be a split infinitive. These make some readers wince, while others can't see what the fuss is about. In many cases, the problem can be eliminated by moving the offending word elsewhere in the sentence, as in "It is rare to bring a project in on time easily". However, this often means that the sense starts to get a bit confused. The best rule is to avoid splitting infinitives unless you feel that it is necessary to keep the sense, clarity or rhythm. Consider:

> Hackitout Software wishes to more than double its turnover in the next financial year.

Without a wholesale reconstruction, the split infinitive is essential here – and what's wrong with it anyway?

Don't end a sentence with a preposition

Prepositions are words like *at, by, from, in, on, down, to, up, for, of* and *with*, which are used to connect nouns to verbs and words to sentences. As the name suggests, in strict English they should go before the word they modify, although in many cases this looks overly pedantic. Would you say, "What's this widget for?" or "For what is this widget?"

Ending a sentence with a preposition sometimes looks clumsy, as in this example:

> No requirements have been discussed for permissioning of metadata, however it is assumed that it is desirable for users to have access to only the metadata for data which they have some level of access to.

But sometimes it is fine, as here:

> The system took three hours to set up.

In that last example, the suffix *up* makes *set* into new verb with a different meaning. People learning English find such things very troublesome: *set up*, *set down*, *set aside*, *set against* and *set on* are all very different. And not all combinations are valid – for instance, *set with* is not a single construction. In the next example, we have another invalid combination:

> The field is not wide enough to allow long surnames to be updated in.

If the incorrect construction "updated in" were changed to "filled in" then the sentence would be more acceptable because *fill in* is a valid combination.

The two components must always occur in the right order, and you should not split these elements up. Why does that last sentence look ugly? Because *split up* is a single construction whose two components have drifted apart. Now what about this one?

> If any feature of Microsoft Word's grammar checker is annoying then you can just turn that particular type of check off.

It looks OK, but the central verb is *turn off*, not *check off*, and its components have been separated. Better is:

> If any feature of Microsoft Word's grammar checker is annoying then you can just turn off that particular type of check.

So, when a multi-word verb is left intact, you can use it to end with. Why does that last sentence look wrong? Because *end with* is a not a single construction but the normal use of *with* – and we expect something on *both* sides of that word. It is these situations where you are left dangling, looking for more words, that are to be avoided. Here is another example:

> IT Consulting is an area that Global Sourcing is currently working with the business on.

The ending leaves you asking, "on what?" Only by going back and re-reading does some sort of sense emerge.

The grammar checker in Microsoft Word will flag some end-of-sentence prepositions. If you can find a better ordering for the words then do so, otherwise the flag can be ignored.

Rules you should remember

So those are the rules you can dispense with. In this section, I will examine the rules that you cannot ignore, which are:

- Make the subject agree with the verb.
- Keep sex out of it.
- Know what your words mean.
- Avoid foreign words and phrases.
- Disentangle 'noun knots'.
- Use the correct plural forms.
- Avoid ambiguity.

Make the subject agree with the verb

For a change, I'll structure this section as a test. Which of the following sentences are correct?

1. None of the suggested solutions provide a complete answer to all the problems identified.
2. Only one of these hardware suppliers are able to deliver a system in the time demanded.
3. He is one of the most experienced analysts who has ever worked in this field.
4. Inter-departmental politics have been the main cause of delay to this project.
5. Neither the Project Manager nor the Divisional Manager have approved any increase to the timescales.
6. Both the overall design and the individual subroutines needs a complete quality review.
7. A large percentage of the programmers have no experience with object-oriented techniques.
8. Hackitout Software undertake to supply the hardware listed in Appendix A.
9. Three years are the normal warranty period for spare parts.
10. Optical media is used for backup storage only when high volumes are involved.
11. The team disagree about the expected end-dates.
12. The Chairman of the Board, together with the Deputy Chairman, are coming to the meeting.
13. Each have a room to themselves.
14. Thousands of pounds' worth of damage were done to the computer equipment.
15. The number coming today are too many for the conference room.
16. We discovered that there were a dozen routines with serious bugs.
17. Bread and butter were all that remained by the time I arrived.

My answers are as follows:

1. Wrong – in this case, the word *none* is singular so it should be "none .... provides".
2. Wrong – the verb refers to only one person, so "Only one supplier . . . is able . . . "
3. Wrong – the verb refers to the group, so "He is one of the most experienced analysts who have ever worked . . . "
4. Wrong – the word *politics* is singular, even though it ends with an s: "Inter-departmental politics has been the main cause . . . "
5. Wrong – "Neither . . . nor", "either . . . or" and "not only . . . but also" all take a singular verb with singular subjects.
6. Wrong – if you have a singular *and* a plural then the verb corresponds to the item nearer to it.
7. Wrong – The "large percentage" is singular even though it encompasses several programmers.
8. Possible – most collective bodies can be singular *or* plural, as in "The committee decides (decide) on Wednesday", "Microsoft are (is) supporting Java". Some company standards dictate that their own name is always singular.
9. Wrong – the time-period is considered a single unit. The same is true in "Six pounds is the joining fee" or "Five megabytes is all the memory we have".
10. Wrong – "media" is already a plural (of medium), so you should say either "Optical media are" or "An optical medium is". The same problem arises with words like *phenomena* and *criteria*, but not with *data* and *agenda*, which are now acceptable as singulars.
11. Right – in general, collective bodies can be singular *or* plural, as in sentence 8. If the stress is on the individual members, the plural verb is better. But if you were to say, "The team has been told to complete their work by Friday", then the individuals are unimportant and the singular is to be preferred.
12. Wrong – although there are two people coming to the meeting, the subject of the sentence, "The Chairman of the Board", is what affects the verb.
13. Wrong – when *each* is the subject of a sentence, the verb and any pronouns should be singular. So the sentence could read, "Each has a room to himself". However, this breaks the sexism rules. Better would be "They have a room each", where the subject is "they" and so takes the plural verb-form.
14. Wrong – the subject was the damage, so "damage was done" is correct.
15. Wrong – when *number* is used as a subject, it takes a singular verb, so "is too many" is correct. However, if you were to say "There are a large number of outstanding orders" then *number* is being used as a part of a plural subject "orders", so there should be a plural verb. Anyway, "a large number of" is a windy phrase – what's wrong with *many*?
16. Right – although *dozen* is singular, it takes a plural verb.
17. Possible – but overly pedantic (you can't win!). Some compounds, usually clichés, are so indivisible that they take the singular verb – like these:

> Give and take is (are?) the key to a happy working relationship.

> The long and the short of it is (are?) that the disks will have to be replaced.

Keep sex out of it

At one time, male pronouns doubled as neutral ones, so *he* could refer to someone of either sex. Now, the language has evolved, so the following examples make sexist assumptions that will offend some readers:

> A child may be in need of compulsory measures or care if any of the following conditions is satisfied with respect to him:
>
> a. He is beyond the control of his parents; or
> b. He is falling into bad associations or is exposed to moral danger; or
> c. The child, being a female, is a member of the same household as a female in respect of whom . . .

> The entire batch of orders will be input by one of the data entry girls. At the start of each batch, she will enter her initials, the batch code number and the batch type code . . . All the MIS reports can be invoked at the designated workstations, the reports being printed at the high-speed printer or at the user's individual printer according to how he has configured his workstation.

> In today's competitive market an IT manager has to ensure that there is a precise fit between resource requirements and availability. Not only must he plan for long-term growth but he must also meet short-term requirements, often at short notice.

You can avoid sex-specific terms without resorting to long-winded, ugly prose, or inventing new words. Somehow, in struggling for correctness, the following example manages to draw attention to the very thing it is trying to avoid:

> If the transaction cannot be automatically entered then it will be submitted to the transaction repair manager(ess), who will be notified by an audible signal at his/her screen. (S)he can then scroll down a list of failed transactions to select the next one that (s)he wishes to correct.

To avoid sex-specific constructions there are a number of possibilities, listed here in rough order of preference:

1. Re-word sentences that employ possessive pronouns, such as *his* and *hers*, for example:

> We will employ a consultant, and after we have read his report we will reach a decision.

Better would be:

> We will employ a consultant to produce a report and, after we have read this, we will reach a decision.

2. Use plural pronouns, for example:

> On hearing the alarm, each user should shut down his machine.

Each candidate must submit his/her application by 12 September, including his/her qualifications and reasons for wanting the post.

Better are:

On hearing the alarm, the users should shut down their machines.

Candidates must submit their applications by 12 September, including their qualifications and reasons for wanting the post.

However, don't employ a plural where the subject is clearly singular, as here:

If the transaction cannot be automatically entered then it will be submitted to the transaction repair operator, who will be notified by an audible signal at their screen. They can then scroll down a list of failed transactions to select the next one that they wish to correct.

One day, this usage may be acceptable, in the same way that *you* became preferred to *thou*. But that hasn't happened yet. In the last example, "They" should be replaced by "The operator".

3. Use the word *you* – for example:

If the user wants to save the file with a new name, he should select the "Save As" option.

This should be written as:

If you want to save the file with a new name, you should select the "Save As" option.

4. Use neutral words to replace sex-specific words. Instead of *businessmen*, *mankind*, *mans*, *man-made*, or *manpower*, write *business people*, *humanity*, *operates*, *synthetic*, or *personnel*. Don't draw attention to the sex of a person doing a job, as in *female data-entry clerk*. Use of words like *manageress* may imply that there is something unusual about a woman holding the role, so it is better to use the neutral version, *manager*. Terms newly minted as sex-neutral, such as *chairperson*, may look over-pedantic in the sort of documents we are talking about, but that is a personal choice.

5. Use *she* in the places where you could use *he*. In my copy of *Access 97 for Dummies*, I read:

If you're convinced that a multi-field key is the answer to your problems, collar your local database guru and ask for her opinion. Hopefully, she can come up with a better solution for the table.

There's no reason why such a guru can't be female and the use of *she* may help balance out unavoidable uses of *he* elsewhere.

6. Use "his or her", "him or her" or equivalent constructions, as here:

At the start of each batch, the data-entry clerk will enter his or her initials, the batch code number and the batch type code . . .

7. Use (s)he, s/he, he/she, manager(ess), or similar bastardised formations only when they are unavoidable.

As a more general rule, you should avoid sexual or cultural assumptions in all examples, artwork, role descriptions and so on. Avoid depersonalising terms such as "the blind"; instead, try to focus on individuals, such as "customers who are blind, or who have low vision".

Know what your words mean

We cannot rely on the word processor to detect when one perfectly valid word is being used in place of another. In the following examples, the writers show themselves either ignorant or confused about the words they are trying to employ.

> While walking down the stairs, I came across a variety of items that should not be stored on the fire escape, as this is the Landlord's demise.

> There has been a mixed response from the four Electronic Document Management System vendors to our Request For Proposal. We have now notified the vendors of our change in tact, and the longer timeframe for evaluation.

> Last year, MegaCorp spent over half a million dollars on corporate merchandise. In some cases, these items have a roll; however, they often have little to offer our sales activities.

> I am interested in some of the free downloads that you advertise on your sight.

The number of possibilities for word confusion is almost infinite. Table 6.1 shows some of the most common.

All writers should have a dictionary on they're desk. If you have any doubts about a word, it only takes a moment to look it up, or to check it with the thesaurus in Microsoft Word. This is a moment well spent if it prevents you appearing illiterate. But, as always, the problem lies not with what we don't know, but with the faults in what we *do* know. If you misused a word in the first place then you won't spot the error when you review that page. So get as many other people as possible to read what you have written – more eyes means fewer mistakes. And yes, I did make a deliberate mistake in this paragraph – did you spot it?

Which and that

One pair of words I could have included in Table 6.1 is *which* and *that*. The grammar checker in Microsoft Word will flag the *which* in the following sentence as incorrect:

> Transactions which require immediate processing will be stored locally.

TABLE 6.1. Easily confused words					
The word . . .	Is not the same as . . .	The word . . .	Is not the same as . . .	The word . . .	Is not the same as . . .
Ability	Capacity	Convince	Persuade	Foreword	Forward
Abuse	Misuse	Council	Counsel	Historic	Historical
Accent	Ascent, or Assent	Credible	Creditable, or	Illegal	Illegitimate,
Accept	Except		Credulous		or Illicit
Adverse	Averse	Defective	Deficient	Include	Comprise
Advice	Advise, or Inform	Definite	Definitive	Infer	Imply
Affect	Effect	Dependant	Dependent	Insure	Ensure
Allude	Elude	Discreet	Discrete	Intense	Intensive
Allusion	Delusion, or	Distinct	Distinctive	Later	Latter
	Illusion	Economic	Economical	Lay	Lie
Alternate(ly)	Alternative(ly)	Effective	Effectual,	Maybe	May be
Amend	Emend		Efficacious,	Militate	Mitigate
Amiable	Amicable		or Efficient	Observance	Observation
Among	Between			Oral	Verbal
Assure	Ensure, or Insure	Elapse	Lapse	Practicable	Practical
Because of	Due to	Elicit	Illicit	Practice	Practise
Beside	Besides	Elusive	Illusory	Precedence	Precedents
Cite	Sight, or Site	Eminent	Imminent	Presently	At present
Complement	Compliment, or	Envelop	Envelope	Principal	Principle
	Supplement	Equable	Equitable	Recourse	Resource
Compose	Comprise	Especially	Specially	Respectfully	Respectively
Comprehensible	Comprehensive	Exceeding	Excessive	Stationery	Stationary
Continual	Continuous	Farther	Further	Their	There, or
Continuance	Continuation,	Fewer	Less		They're
	or Continuity	Flaunt	Flout		
		Forever	For ever		

This is because we are making a *restriction* on the type of "transaction"; there are other types not requiring "immediate processing". Purist grammar rules state that when you make such a restriction, you should use *that* rather than *which*. However, Word will not flag the following as an error:

Alerts, which require immediate attention, will be sent to the operator's console.

Here, we are not restricting the "alerts" but describing a property common to them all. In these circumstances *which* is correct.

Personally, I doubt whether 99% of readers would spot this distinction, and it's not something that (which?) gets my goat. In fact, I often find that Microsoft Word has flagged my incorrect *which* clauses, which (that?) it spots by looking for a comma immediately before the offending word. However, there are some cases where care is needed to avoid ambiguity, as here:

Re-partitioning is a form of defragmentation, which is always a risky process.

Re-partitioning is a form of defragmentation that is always a risky process.

As, for and because

The words *as* and *for* cannot be used in place of *because*. In the following examples, *because* would make the sense clearer:

The project is five weeks late as the integration tests revealed several problems with host sessions dropping out.

We have no need to order extra disk drives for the current volumes have not filled up as quickly as predicted.

Avoid foreign words and phrases

The rule that you shouldn't use words that you don't understand is particularly applicable to foreign phrases. A desire to appear sophisticated can leave you exposed to ridicule, as these examples show:

The workstation is required to record all messages received so that a full reconciliation can be made by Securities Management between instructions received viz. a viz. those printed and recorded in the MegaCorp system.

Once the operator's bona fides have been checked by the Security server, the log-on process is completed by displaying the initial menu screen.

A Data Warehouse solution may not always be apropos to the demands of real-time transaction processing applications.

So be careful with the phrases shown in Table 6.2 and always consider choosing the English version first.

It is unnecessary to use *re* when heading letters, as in "Re: Our Order for a Sun Workstation". Just leave it out.

As per is a clumsy multilingual formation that tends to be used by the pompous, as shown by this example:

The modules have all been tested as per the Specification of 27 July.

The *as per* should be replaced with *in accordance with* or *as in*.

	TABLE 6.2. Foreign phrases		
Phrase	**English equivalent or meaning**	**Phrase**	**English equivalent or meaning**
Ad hoc	One-off, Special-purpose, Improvised	Laissez-faire	Live and let live
		Modus operandi	Way of working
Apropos	With respect to		
Bona fide	Genuine, In good faith	Mutatis mutandis	With the necessary changes
Carte blanche	Freedom, A free hand		
		Per annum	Yearly
cf	Compare	Per capita	Each
Circa	About, Around	Per diem	Daily
De facto	Established by common practice	Per se	As such, Essentially
De jure	Established officially	Pro forma	Blank, Sample, Template
e.g.	Such as, For example	q.v.	Which see
En masse	Together, As a whole	Re	About, Regarding
Etc.	And others, And the rest , … , For example, And so on	Versus, vs., v.	Against
		Via	By means of, By way of
Ibid.	In the same place	Vis-à-vis	Regarding, As regards, About
i.e.	That is, Which is		
Inter alia	Among other things	Viz.	Namely

Etc.

Special mention must be made of *etc.*, the lazy writer's ally. I always regard *etc.* as meaning "I couldn't think of any other examples", as here:

> *Writing to Win* provides some tips for improving the quality of our proposals, reports etc.

You have to be honest. If you really can't think of any more examples then just miss out the *etc.* If you *can* think of some then it is better to precede the list with *for example* or *such as* and so remove the suspicion that you might have less ammunition than you pretend. So the following would be better:

> *Writing to Win* provides some tips for improving the quality of our business documents, such as proposals and reports.

Avoid having *e.g.* and *etc.* in the same sentence. In the following example, the commitment is deliberately vague, but the fudge is overcooked:

> We will provide a full documentation set, e.g. program listings, user guides (e.g. help text), technical manuals (e.g. supplier manuals) etc.

Disentangle noun knots

Consider the phrase "frequency-shift power-line carrier relay module". The words *frequency*, *shift*, *power*, *line*, *carrier* and *relay* are normally nouns, but are here employed as adjectives to specify a certain type of "module". Using one or two nouns in this way is a perfectly valid use of English: *disk drive, England football team* and *Citizens' Advice Bureau* are some examples. But as the number of nouns rises, the expression becomes harder to unravel, and the reader is slowed down. I recently received a document entitled:

Third Party Real-Time News Collection Unification System Test Plan

I had little interest in reading further.

The key to disentangling a noun knot is to convert one of the nouns into a verb. In the example of the test plan, the principal function of the document is to describe testing. What is being tested? A "system". So the title should start with "Testing the System" – a verb and its object. That system can then be described in more detail by clauses that follow those important words. The result is:

Testing the System That Unifies the Collection of Real-Time News Obtained from Third Parties.

It is useful to identify when you are using a noun knot rather than a list of adjectives because they have different punctuation rules. The test is as follows:

1. Insert an *and* between each pair of words.
2. For the cases where that seems to work, you have two adjectives. Replace the *and* with a comma or a space (see the rules for commas in the chapter on punctuation).
3. Otherwise, try a hyphen between the words.
4. If that seems to work then remove the hyphen – you have a noun knot.
5. If neither the *and* nor the hyphen looks right, you are at a junction between a list of adjectives and a list of nouns. Check that all the adjectives come before the noun knot.

Let's try the test on this sentence:

After four months of testing, all that has been delivered is an unusable unreliable sales analysis application

The last phrase comes out like this:

. . . an unusable and unreliable sales-analysis-application.

So *unusable* and *unreliable* are adjectives and should be separated by a comma, while *sales analysis application* is a noun knot and does not need the hyphens. The result is as follows:

> After four months of testing, all that has been delivered is an unusable, unreliable sales analysis application.

Use the correct plural forms

The following examples demonstrate confusion about plural words, especially foreign ones.

> Appendix A contains the Curriculum Vitae's for the proposed project team.

> We first set out a number of criteriae for the selection of a suitable product.

> The new telephone-based bill payment service enables you to pay telephone, mobile phone, credit card, electricity, insurance premia and other utility bills over the phone.

Table 6.3 shows the correct plurals of some troublesome words:

> The strict Latin plural of *status* is *status*, pronounced **stay**-tews. I would tolerate *statuses*, given the awkwardness of the situation, but there is no excuse for *statii*.

Singular and plural

There are times when you wish to indicate that both the singular *and* the plural of a word are applicable, as here:

TABLE 6.3. Troublesome plurals			
Word	**Plural**	**Word**	**Plural**
Agendum	Agenda	Formula	Formulae
Album	Albums	Index	Indices
Apex	Apices or apexes	Memorandum	Memoranda
Apparatus	Apparatus	Parenthesis	Parentheses
Appendix	Appendices	Premium	Premiums
Basis	Bases	Prospectus	Prospectuses
Crisis	Crises	Species	Species
Criterion	Criteria	Status	Statuses (see above)
Curriculum vitae	Curricula vitae	Virus	Viruses

> Analytics are product-independent algorithms that take data input(s) and derive output(s). The output(s) are assumed to be owned by MegaCorp and are independent of delivery, desktop, or host.

If possible, you should reword, giving the following:

> Analytics are product-independent algorithms that take one or more data inputs and derive one or more outputs. The output, or outputs, are assumed to be owned by MegaCorp and are independent of delivery, desktop and host.

If you feel that the rewording is too cumbersome then the (s) construction must be used instead. The result should be treated as a plural, so don't say "The output(s) is/are assumed … "

Words with irregular plurals do not take happily to this treatment, as shown by this example:

> A Domain consists of one or more capabilities, as described in its *Domain Architecture* Document. The constituent capability(ies) are then further defined in their own *Capability Architecture* documents.

Rewording is then the only option. Here, the phrase "capability or capabilities" would suffice.

Possessives and plurals

Make sure that you can distinguish between a plural and a possessive, which needs an apostrophe-s formation. The writers of these examples have not done so:

> Use Case Points can be calculated at a much earlier stage in the projects life.

> The latest version of Microsoft Project will allow task estimates to be imported from Excel spreadsheet's.

I cover the use of apostrophes in the chapter on punctuation.

Avoid ambiguity

There are two reasons why you should not allow a sentence to be ambiguous. One is that it could give the reader a misleading impression, which might not be to your advantage. The other is that it may cause the reader to come to a halt, wondering how the sentence is to be interpreted. We never want our readers to study the text itself; the message we are trying to convey should be transmitted without any such hiccups. There are several ways in which ambiguity can arise – here are some of them:

- Where a pronoun, such as *it, he* or *she*, can refer to either of two previous terms, as here:

The Architecture Group have authorised the purchase of a new accounts package. Tomorrow it will be installed and soak-tested.

Is it the Architecture Group or the accounts package that will be installed?

- Where part of a sentence may refer to one or more previous parts, as in this example.

As the report recommended, we have looked into the problem and we made a number of improvements.

Were the improvements recommended by the report or not?

- Where there is a list of items described by a single adjective or adverb, as in these sentences:

The report includes coloured organisation charts and data flow diagrams.

The committee methodically evaluated and costed the proposals.

Were the data flow diagrams coloured? Was the costing of the proposals methodical?

- Where a negative may apply to more than one term, as here:

The device has no moving parts and illuminated buttons.

Does the device have any illuminated buttons?

- Where punctuation fails to make the meaning clear, as in this example:

The program ran until the first routine where there was some screen output.

Did it stop during the very first routine, or did it run until there was some screen output?

- Where the use of numbers has been oversimplified, as here:

For the third week in succession, a senior manager has been unable to access the sales figures.

Was it the same manager each time?

- Where words have been omitted, as in this sentence:

The security guards were ordered to stop sleeping overnight in the office.

Was it the guards or other office workers who were sleeping?

- Where the sentence is phrased in the wrong order, as in this example:

  The test team will take over solving problems from the development team.

  Is it the development team that has problems, or the programs they produce?

- Where a word has multiple meanings, as here:

  A junior programmer has been suspended over the missing files.

  Is the programmer in mid-air?

- When a metaphor is being used, as in this sentence:

  Wait until our backs are up against the wall. That's when we'll turn around and start to fight.

  But won't we then be facing the wall rather than our enemies?

CONCLUSION

There is a lot more I could say about grammar, but this chapter has covered the rules that are broken most often, and I suspect that you have had enough. There are plenty of books that cover such issues in more depth – see Appendix B (Select bibliography and resources).

So how can we avoid issuing documents that contain the clumsy, inept English that devalues our message? Firstly, we must constantly be aware of the traps – the type of constructions that might introduce error or ambiguity. It doesn't matter if something has to be explained in a more long-winded manner if this will make it clearer. Try re-reading each paragraph as if you had never seen it before. Does it flow? Is it too clever? Can it be interpreted in any other way? Secondly, you must get other people to review your work. I have made this point before, and I will do so again. Don't be afraid to criticise or to take criticism, for this is the only way that your proposal can be 'debugged'.

CHAPTER 7

Obeying the punctuation rules

THE PURPOSE OF PUNCTUATION

The subject of punctuation can raise considerable debate. One camp maintains that it helps the reader in the same way as musical notation helps an instrumentalist to interpret the notes as the composer intended. The other camp maintains that a fussy succession of little marks interrupts the flow of information to the brain, and that punctuation is only needed when there is a possibility of ambiguity. But there is a difference between those who mis-punctuate through ignorance and those who carefully choose when to omit unnecessary marks. When I read reports and proposals, I often sense that writers are scared to use anything more than the minimum of punctuation in case they get it wrong. And then they *do* get it wrong. The effect is to devalue the points being made because the text is so hard to unravel, as the following example illustrates:

> The data is integrated in the database and any user interface screen dealing with some data will have to access the database to pre-populate with any data that has already been entered – so we need notification and work scheduling mechanisms so that the (for example the Journalists and Data Analysts can co-ordinate their effort, and that they can be notified when other relevant input occurs such as reporting information on exchange feeds).

You *can* go too far in the other direction, littering your text with so many marks that reading it becomes a series of fits and starts. But, in general, I side with the "musical notation" camp. Commas, full stops, dashes and the rest of the punctuation armoury help marshal your text into readable chunks. If they are omitted, or inserted at the whim of the writer rather than to help the reader, then the prose becomes harder to understand and your readers get bored. Effective punctuation will ensure that your message is clear and unambiguous – and we don't want to lose any opportunities to do that.

So the aimX of this chapter is provide a refresher course and to encourage you to make more use of the right marks. And even if you think that punctuation is old-fashioned and pedantic then remember that there are plenty of people out there who don't – and who may bin your proposal at the first sight of a misused apostrophe. Even if you choose not to follow all the rules, it is useful to know what they are.

STOPS

A 'stop' is the generic word for any punctuation that brings your reading to a temporary halt or pause. The term embraces full stops, commas, semicolons, colons, dashes, brackets, question marks and exclamation marks. In this section, I will discuss all of these and show where they should and should not be used.

Stops have two purposes:

- To divide an argument into individual points and to show how each point should be read and interpreted.
- To prevent ambiguity. See how I can change the sense of the following sentence by re-punctuating it in different ways.

> BigCo has one major problem: only we know the root password.
> BigCo has one major problem only: we know the root password.
> BigCo has one major problem only, we know: the root password.
> BigCo has one major problem only we know: the root password.

So if we use the right stops then the reader is left in no doubt about what we are trying to say – and that is a goal worth the effort of a little study.

Full stops

I have already discussed sentence length in the chapter on grammar. You want to aim for a mixture of short and long sentences with an average of around 15 to 20 words in each.

A sentence expresses a single statement, so don't be tempted to append additional points, gluing them on with conjunctions, such as *and* or *plus*, or with commas, as the writers of these examples have done:

> The *Getting Started* task pane in Microsoft Project is similar to the task panes you see in other Office applications plus it provides a convenient list of recently opened files as well as an additional method of creating new files.

BigCo has released a new version of the MAGUS financial accounting package, their rivals are soon expected to match the facilities it offers.

In both these cases, the writer has not concluded their first point before running on into a second. Either a full stop or a semicolon should be used, depending on how closely the two points are connected.

Commas

Use of commas is one of the first things I analyse when I feel that something I have written is not coming over as well as it should. By breaking up the sentence, I can often see if the individual little clauses make sense in themselves, if they can be strengthened by slight changes to their wording or if I can change the way they are connected.

There are some fixed rules for the use of commas, and some areas where their use is a matter of taste. Four main usages will cover most of the cases liable to be met in business writing. These are:

- To divide lists of words or phrases.
- To enclose subsidiary information.
- To close the introduction of a sentence.
- To connect parts of a sentence together.

I will discuss each of these in turn.

Dividing lists

There are four circumstances where a comma is used to divide the items in a list:

1. To substitute for the word *and* in a list of the form 'A, B and C' or for the word *or* in a list of the form 'A, B or C'. For example:

   Hackitout's Integrated Support System provides compatibility between tools and workers, consistent application interfaces, ease of learning, user friendliness and expandability.

   In that example, the last item in the list ("expandability") is preceded by an *and*, making it a list of attributes. If the last item is indicated with an *or* then we have a list of alternatives. Either way, the commas prevent the need to put an *and* or an *or* between every item in the list.

   Whether a comma should precede the final *and* or *or* is a matter of taste. In the UK, the convention is to omit this comma, while in the USA one is usually inserted. Microsoft Word provides an option that allows your local convention to be checked. If there is any possibility of ambiguity, which can occur when the items in the list are long or contain the word *and*, then the final comma never does any harm.

2. In an **adjectives list**. This occurs when two or more adjectives are being used to describe something. For example:

> This configuration features a reliable, load-sharing, multiprocessor solution.

Again, the comma replaces the word *and*.

    Where the adjectives are being used to convey a single idea, rather than the individual qualities of something, there should be no commas between them. For example:

> The CPU is housed in a massive shiny black cabinet.

3. To separate items in dates, addresses and titles when they are 'in line' in the text, for example:

> The proposal was finally signed on June 20, 1998.

> Address all complaints to Gordon Bennett, The Chief Fobber Off, BigCo Ltd, Stevenage, Hertfordshire.

The current convention in the UK is to omit the commas in the address at the top of a letter or on an envelope, after "Dear Mr X" and after closures such as "Yours sincerely". They are usually retained in the USA.

4. To separate the thousands digits in numbers other than dates, as here:

> The total cost of the system at 2003 prices is £1,347,900.

Enclosing subsidiary information

A pair of commas is used to enclose any sort of comment, expansion or explanation that interrupts the sentence. If you remove the phrase in the commas, the sentence should still make sense, as in these examples:

> Software estimation, which is still considered a 'black art' in some quarters, can be tackled using a methodological approach.

> We will, therefore, continue with the project.

You can also use dashes or brackets for this purpose, brackets being less forceful than commas and dashes more so, as shown here:

> The final and most complex exchange adaptor, London, was completed in May

The final and most complex exchange adaptor – London – was completed in May.

The final and most complex exchange adaptor (London) was completed in May.

Don't forget one of the commas enclosing the interruption to the sentence, as the following writers have done:

The system you will be pleased to hear, is now running perfectly.

The CEO of MegaCorp, Martin Smith is visiting the project team on Friday.

Finally, don't use these 'bracketing' commas unless the phrase within them can be removed without destroying the grammar or the meaning of the sentence. Look at these examples:

A laser printer, which will do a better job, and be more reliable costs less than £200.

People, who live in glass houses, shouldn't throw stones.

In the first case, removing the phrase in commas leaves an invalid sentence; the second comma should be placed after "reliable". In the second example, removing the phrase "who live in glass houses" destroys the whole point, so it should not be placed within commas.

Closing the introduction to a sentence

A comma should always follow *accordingly, consequently, for example, for instance, further, furthermore, however, indeed, moreover, nevertheless, nonetheless, on the contrary, on the other hand, also, thus* and similar introductory words. An introduction can be one of these or a longer phrase, as shown by these examples:

However, most people find estimating software effort a more daunting task than sizing hardware systems.

For example, not many computer games are targeted at the over-60s.

During software engineering projects, it is taken for granted that serious changes to the functionality can be introduced at any stage.

Although we are new to facilities management, our previous work with MegaCorp has given us seven years of background experience.

An individual clause within a sentence may have its own introductory phrase, as here:

Our revenue was down during the third quarter, but, when we analysed the figures, we realised this was caused by staff holidays.

Not every sentence has an introductory phase, as these examples show:

One way to keep your schedule up to date is to record what actually happens to each task in your project.

The best people to undertake IT project estimates are those who are going to undertake the work.

Familiarity of technical writers with computer languages often leads them to assume that every *if* needs a corresponding *then*, as in this example:

If a project loses a million dollars, then it might take ten million dollars worth of successful work to regain the break-even point.

However, you don't need a comma and *then*. Usually just the comma will do. Reserve *then* for cases where the corresponding *if* is a long way away, where there is a possibility of ambiguity or where you particularly wish to stress that one action is a consequence of another.

Connecting the parts of a sentence

Commas should precede the break points in a longer sentence. These often start with words like *and, or, but, so, yet, which, where, although, as, while* and *also*. However, there are many other ways in which a new clause can begin, and there is no universal rule that can detect these. Some people find it easiest to think of these situations as the places where you might pause for breath while reading the sentence aloud. If you are alone, you might like to try that on these examples.

Hackitout Software wishes to continue with the original contract, but BigCo intends to walk out and take the matter to court.

On this project, the Design Team Leader will be Fred Smith, who prepared the original Functional Specification.

Like other Office applications, Project customises the menus and toolbars for you, based on how frequently you use specific commands or toolbar buttons.

During software engineering projects, it is taken for granted that serious changes to the functionality can be introduced at any stage, while still maintaining the same timescale.

Another example of where you might pause for breath is just before reporting speech or quotations, as here:

Our CEO said in 1999, "MegaCorp must truly embrace the Internet".

Bad uses of commas

If you learn the four main usages for commas that I have described, and apply them rigorously, then you will be communicating better than most of the people who write reports and proposals for a living. However, I just want to mention five cases where using a comma actually makes a sentence less clear.

1. *Because* is not usually preceded by a comma. It breaks the flow and implies you are starting a new point, as in this example.

   Two of the test team are currently idle, because the Java code modules are not being developed quickly enough.

2. Words and phrases like *accordingly, consequently, further, furthermore, however, indeed, moreover, nevertheless, nonetheless, on the contrary, on the other hand* and *thus* are used to start new sentences, so cannot be preceded with a comma. In the following example, the word "however" should be the start of a new sentence (and followed by a comma).

   All software companies pay lip service to a defined methodology, however Hackitout Software's CONTRIK programme is built in to all our operational procedures.

   On the other hand, such words and phrases can be used within pairs of bracketing commas, as here:

   Outsourcing, however, removes many day-to-day worries, replacing them with one big worry every time the contract is renewed.

3. Other than the specialised use in lists that I have described, a comma cannot replace a conjunction. In this example, the comma needs to be followed by the word *but*, or replaced by a semicolon.

   The printer toner cartridge has run out, it will be renewed.

4. In some sentences, conjunctions do not need a comma before them. This is when the two parts of the sentence are short, have the same subject or introductory phrase, or are otherwise closely connected. For example, there is no 'pause for breath' before the *and* or *but* in these sentences:

   The conference will continue all afternoon and into the evening.

   The user interface is functionally rich but uncluttered.

   I prefer to minimise my office time and work alone.

   This is where we reach the boundary of personal taste. If you are undecided as to whether a comma is needed, just ask yourself whether its inclusion makes the sentence clearer. It's better to make sure of that than to risk criticism for including some redundant punctuation.

5. By applying *all* the rules, it is possible to overload a sentence with so many commas that it becomes a stuttering series of clauses, impossible to unravel, as here.

    The model has a simple, generic design, which allows new data, and new types of data, to be added, without changing the basic, but still highly usable, structure.

    In such cases, you must identify the most important breaks in the sentence. You can use 'stronger' punctuation, such as dashes or semicolons, to break the flow, or just omit the commas where they are detracting from the overall clarity. The example could be rewritten as follows:

    The model has a simple, generic design that allows new data – and new types of data – to be added without changing the basic (but still highly usable) structure.

Semicolons

Semicolons are under-used, striking some people as somewhat old-fashioned or pedantic. However, there are many occasions where one is exactly right; it is less curt than a full stop, and allows a connected, but equally important, point to flow from the previous statement.
Take a look at this example:

Augmentation does not create a new annuity record, it enhances an existing annuity by adding further benefits, thereby increasing the liability on the payments system.

The comma after "annuity record" is too weak, but a full stop might lead to confusion about what the "it" is referring to. So a semicolon is ideal. However, this next example is wrong; the second part is not a complete sentence.

There are not many books about the estimation of IT projects; just the one last year.

A good use for semicolons is to replace conjunctions, like *and* and *but*, when they join complete sentences. In the following example, two related but equally important points are being made, so a semicolon would be better than *and*:

In the diagram, shaded packages are outside the scope of the accounting capability and dependencies these may have on other non-accounts capability packages are not shown.

You can also use a semicolon as a 'super-comma' in a long and complicated list, particularly when the individual elements contain commas. However, a bulleted list should also be considered in such cases.

Colons

Colons are used in the following situations:

- To mark the end of the sentence that introduces a bulleted or numbered list. This is discussed further in a later section of this chapter.

- To introduce a list at the end of a sentence, for example:

  Our solution meets all your business requirements: speed of implementation, expand-ability, scalability and high quality.

  This format removes the need for words like *such as* or *including*.

- To introduce a specific example or detail. The colon acts as a sort of drum-roll, as in this example:

  Our solution to all these problems is a simple one: a Proof of Concept exercise.

- To act as a divider between two contrasting elements, for example:

  A smaller system would be cheaper and quicker to deliver: a larger one would offer better performance and expandability.

Colons are also used to separate the title and subtitle of a book and in expressing mathematical ratios. The mark of a colon followed by a dash (:–) is extinct.

In the UK, the phrase following the colon does not begin with a capital letter, but in the USA it does, so long as it is a complete sentence, for example:

  Select an option from the left-hand panel: The "Start" button will clear the board for a new game, and the "Exit" button will stop the application.

Dashes

A dash is not the same as a hyphen. The hyphen is used to join compound words and to split words between lines. The dash is a punctuation mark used to indicate the start of an aside, explanation or addition to a sentence. It provides the ability to flow another idea on to the end of a sentence – or into the middle of one – without using a comma or semicolon.

  The operators are demotivated and slipshod – as anyone would be when trying to use such a system.

  Two of Hackitout Software's major Divisions – Defence and Finance – will combine their forces during the design phase.

Strictly, there are two types of dash:

- The **en rule** is as wide as a capital N (–) and is obtained by typing *<Ctrl><Numeric Key Pad Minus>* in Microsoft Word.
- The **em rule** is as wide as a capital M (—) and is obtained by typing *<Ctrl><Alt><Numeric Key Pad Minus>* in Microsoft Word.

I only use the en rule, although some say that this should be reserved for a relationship or a range, as in *client–server system* or *pages 23–25*, and that the em rule should be used for extending sentences.

If you are using a dash to indicate a range then don't write something like:

> The system is running at between 95%–97% of its total capacity.

The idea of *between* is already implied by the dash, so either remove the *between* or say "... between 95% and 97% of its total capacity".

So when should you use a dash instead of a comma, colon or semicolon? It is very tempting to select the dash because it is rarely completely wrong. However, dashes are much 'stronger' than the other marks. If you use them at every opportunity then you are depriving yourself of a way of controlling the rhythm and intonation of your prose. It's a little like shouting all the time – after a while, nobody takes any notice. Compare these three pairs:

> The user interface is not the best in the world, as anyone can see, but the system has been in use for nearly twenty years.

> The user interface is not the best in the world – as anyone can see – but the system has been in use for nearly twenty years.

> System rollout will mark a new beginning; it denotes the renaissance of MegaCorp as a world player in this market.

> System rollout will mark a new beginning – it denotes the renaissance of MegaCorp as a world player in this market.

> BigCo needs just one big project to remain solvent: the Ministry of Defence submarine contract.

> BigCo needs just one big project to remain solvent – the Ministry of Defence submarine contract.

In all these cases, the dash *could* be used, but the alternative has a little more precision, so it is the better choice. The dash can then be saved for cases where a stronger break is wanted. You might also notice that the dash tends to emphasise the words before it, the semicolon makes the two parts of the sentence equal, and the colon throws the emphasis onto the second part. This may help in deciding which mark to employ.

Parentheses

Parentheses contain explanatory ideas. Commas and dashes are also used for this purpose, as discussed earlier. In general, the text in parentheses should be a single sentence or less, but I have often seen a whole paragraph whispered in this way. Are we supposed to ignore it? Or read it only if it looks interesting? Either such text belongs elsewhere or else it should be omitted.

Avoid having a punctuation mark, then a bracket, and then another punctuation mark. Usually, any final punctuation goes outside the brackets, as here:

> The results prove the bug lies in the Accounts module (probably in the interface to the General Ledger). (See Appendix A for the full test results).

Square brackets are used to add additional words to quotations if they don't quite make sense in the context:

> Bill Gates once said, "640 K is enough [memory] for anyone".

Question marks

Question marks should not be used for indirect questions or for polite requests, as these examples show:

> During our study, we asked users whether the present interface hampered their work.

> Will you please forward your comments before the end of March.

Throwing a question into the text and then answering it is an effective technique, if it is not overdone. For example:

> The conclusion is that replacement of the present system is long overdue. But supposing there is no budget for this until April? The only solution would then be to partition the existing disks, thus sacrificing performance for an increase in capacity.

Constructions like "But why is this?" or "How is this to be achieved?" can replace more conventional joining words like *however* and *therefore*.

A question can also be used for a forceful header, such as this:

Why Upgrade the Processor Now?

Exclamation marks

Exclamation marks are rare in formal prose. There is a danger of seeming too gushing and enthusiastic, especially if you are in the habit of using two or three of them in succession, which is never acceptable in formal prose. In general, you should avoid exclamation marks unless you are making a truly striking point that you don't want to disappear into the mass of text, as here:

> Response times have deteriorated rapidly over the last three months. One user claimed that at peak times it could take over twenty minutes to enter a single transaction! This situation cannot continue.

APOSTROPHES

Misuse of apostrophes is the most common error in the punctuation of business documents. At present, we are in an intermediate phase in the evolution of these troublesome tadpoles. One generation has been educated with all the rules, so misuse jumps out at them. Some of the next generation have undertaken sufficient business writing to understand the rules and apply them grudgingly. The latest generation just stick in an apostrophe when it seems that there hasn't been one for a while. Undoubtedly, the apostrophe is doomed. However, our job is not to assist the evolution of the language by making or condoning mistakes. While there are readers out there who will be critical of misuse, we must ensure that we do not give them cause for complaint.

Possessive apostrophes

Many people emerge from the education machine with only one punctuation rule etched into their consciousness: apostrophe-s. Unfortunately, they think there should be an apostrophe before *every* terminating s. As I sit here now I can see a clothing store advertising that its sale "End's Friday" and a greengrocer displaying a selection of:

> English apple's
> Fresh mush's
> Prime cabbage's

Actually, one of these is correct – can you see which?

The apostrophe-s is only added when a noun *possesses* the next noun mentioned. So in the phrase *the emperor's new clothes*, the emperor is shown to possess the clothes. The full rules for possession are as follows:

1. For the simple case just add an apostrophe-s. Start with the word *cat*; the cat possesses some pyjamas; so if you want to refer to those very pyjamas, you add an apostrophe-s to the cat and get *the cat's pyjamas*.

2. If the possession is shared then only the last item gets an apostrophe, as in "We evaluated Fred and Jim's suggested solution". In that example, replacing *Fred* with *Fred's* could mean that there were two alternative solutions.

3. For plural words, the apostrophe goes after the *s*, as here:

   Several manufacturers' benchmarks show that our products outperform the competition.

4. Words ending in y, which are pluralised with *ies*, do not do so just to take the apostrophe-s. So *the company's property* is stuff belonging to the firm, while *the companies' property* belongs to several organisations.

5. Irregular plurals and acronyms also need the apostrophe-s formation – for example, *the children's games* and *an OEM's products*.

6. Non-plural words that end with an s can take an additional apostrophe-s or just an apostrophe. For example, *Pythagoras' theorem* and *Mr Jones's letter* are both correct. Use whatever *sounds* right. Awkward constructions such as *trousers' pockets* should be avoided by re-phrasing.

7. The same rules apply to compound constructions like *each other's office* or *anyone else's idea*.

8. Products and product features should not be allowed a possessive form. For example, *the Windows' interface* and *the spelling checker's dictionary* are better expressed as *the Windows interface* and *the dictionary in the spelling checker*.

9. Sometimes the object being possessed is not mentioned, but the possessive indicator is still needed, as in these examples:

   MegaCorp's products are good, but BigCo's are better.

   She is at the Doctor's.

10. Finally, the apostrophe-s formation is used to indicate the passage of time – for example, *a month's pay* and *three years' warranty*.

Apostrophes and abbreviations

An apostrophe can be used to indicate one or more missing letters, as in the three cases in this example:

   We don't recommend connecting the blue wire to the earth terminal – that's what caused the comm's to blow last time.

So in the greengrocer's sign I mentioned earlier, *mush's* would be correct if the apostrophe were being used to indicate the missing letters in *mushrooms*, but not because the word is plural. In formal writing, the use of apostrophes to indicate omission should be eliminated; all words should be spelled out in full. You may have noticed that one of the ways I achieve the chatty tone in this book is to use contractions like *don't*. Don't.

Difficult plurals

With some words, it is difficult to form a plural that doesn't look ugly. Sometimes you need to insert an apostrophe, not to indicate possession or omission but just to provide some 'breathing space', as in the following:

> There is a long list of do's and don'ts when choosing between different PC's. Since the 1990's, I have chosen the one with the most A's in the *PC World* rating tables.

As always, you should err on the side of clarity. Personally, I would write "PCs" and "the 1990s". However, a 'breathing space' apostrophe should definitely not be used in cases where you can't be bothered to think of the correct plural, as here:

> Periodically, the system undertakes a bulk update of the PAYE tax code suffix's resulting from the budget.

Apostrophes in pronouns

Now things get difficult. Pronouns are words like *him*, *her*, *it* and *they*. Each one has a possessive equivalent: *his*, *hers*, *its* and *theirs*. The trouble is that these already have possessiveness built in, so they don't need an apostrophe. Few people would think of writing something like this:

> Each consultant will bring hi's own PC if the client cannot supply sufficient of their's.

However, every day brings further examples of a misused *it's*, like these:

> Your application should be approved by your staff consultant, who will advise you of it's approval.

> Technical evaluation: establishing the design objectives of the system, it's stability and maintainability, it's fit to MegaCorp's technology and data requirements, and it's ability to meet required performance levels.

> Hackitout Software – professionalism at it's best!

The problem is that *it's* is a perfectly valid as an abbreviated form of *it is* or *it has*, with the apostrophe indicating that one or more letters have been omitted. So, although those examples may look OK at first sight, many readers will read each *it's* as *it is*, making the text meaningless. And they will then realise that the writer doesn't know the rules of the language, which may weaken the authority of the prose. As I said earlier, the use of an apostrophe to indicate omission should not be undertaken in formal writing. If you're going to get it wrong then at least err with the least embarrassing option, so **don't put an apostrophe in the word _its_**.

There is a similar difference between _who's_ and _whose_, which has not been understood by the writer of this example:

> The decision on the relevance of a particular course, the level of financial or other assistance will be made by the line manager, who's judgement is final.

Who's is short for _who is_ (or _who has_), while _whose_ means 'belonging to the person previously mentioned', and should have been used in the example.

OTHER MARKS

Quotes

There are two uses for quotation marks. The first is when you are quoting something or someone, including yourself, for example:

> All the facilities listed in the Requirements Specification as "essential" will be provided, while those listed as "highly desirable" will be implemented if time allows.

> By "reference data", I mean everything other than real-time and time-series information.

> "Booting up" on page 42 describes how to start the system.

The second is when you are using a word in a special way, such as:

> The task has been 'time boxed', so will be completed by the 16 May.

Personally, I always use double quotes for the first case and single for the second. In the USA, double quotes are the norm for both usages.

When quoting, particularly for longer extracts, you do not have to use quotation marks at all if you set the text apart by using a different margin, typeface or text size.

Whether any punctuation belonging to the 'outlying' sentence should be inside or outside the quotes is mainly a matter of personal taste and local conventions. Microsoft Word has a setting in its grammar checker that will flag exceptions from the chosen standard. I put everything outside the marks unless it _must_ belong with the quotation:

> During our meeting, you referred to a possible "preferential trading relationship", so I would like to present our ideas on this topic.

> Our Systems Architect has stated, "IP everywhere".

> At our introductory meeting, your chairman, Mark Jones, asked, "Are there any low-cost IT projects that will improve MegaCorp's cash flow?" Our proposal answers his question.

In the USA, the convention is to place everything *inside* the quotation marks unless it *must* go outside.

Where there is a potential pile-up of punctuation marks, as in the last example, you can omit some, so long as the sentence still makes sense. Here, the question mark is essential, but it is already a stop, so you don't need an additional full stop after the closing quote mark. This example also illustrates that if the quotation is a complete sentence then it should begin with a capital letter, even if it is embedded within one of your own sentences.

Quotation marks are also sometimes used to introduce a new term, which can then be employed without the quotation marks at subsequent places in the document, as here:

> The framework supports two classes of event: an 'explodable event', which is the original event, and a 'sendable event', which is the resulting action. Explodable events are persistent . . .

I prefer to use bold text for this purpose.

Hyphens

The hyphen is used to join compound words, but only when they are used as adjectives. For example, you may refer to a *look-up table*, a *far-ranging review* or a *low-scoring assessment*. When the compound is used as a noun or a verb then omit the hyphen. For example, say *log on* not *log-on* and *cross section* not *cross-section*. The noun knots discussed in the previous chapter do not incorporate hyphens, so you would refer to a *disk filename extension*.

Compounds beginning with prefixes such as *non-, co-* and *re-* usually retain their hyphens, as in *non-executive* and *co-operation*. Follow dictionary guidelines for special cases like *co-operate, unnumbered* and *misspell*. Notice that such a hyphen can change the meaning of the words, for example, *re-covering* a disk is not the same as *recovering* it.

A hyphen is also used when words must be split between lines. In proposals, this is not usually a problem except within closely packed tables, in which case consider a different type size or font before letting hyphenation intrude.

Hyphens are also used when numbers are written as words: *fifty-nine* rather than *59*. In general, numbers under one hundred should be spelled out. However, if a discussion includes a mixture of numbers under and over this limit then use numerals throughout. Numerals should also be used for page numbers and for percentages. Avoid using them at the start of a sentence.

Slashes

Use of a slash mark to replace the word *or* is a lazy habit; it looks ugly and can be ambiguous. If the alternatives consist of more than one word, there is a risk that the sentence will make no grammatical sense when read one way rather than the other. For example:

> The analyst corrects the Organisation codes on the METOS system following emails/phone calls to/negotiations with TOREP analysts.

> Stories received/phoned through to the news desk and/or from third parties are entered into the Editorial system.

In the second example, the alternative "stories received to the news desk" makes no sense. Also, the proximity of the two different sets of alternatives makes you wonder if there is only *one* set. You end up scanning the sentence in different ways, searching for the different meanings.

Use of *and/or* can always be avoided. Rather than saying

> Commands can be entered using the keyboard and/or the mouse.

go for

> Commands can be entered using the keyboard, the mouse, or both.

The only valid use of a slash mark is within some conventions such as *on/off, CR/LF, read/write* and *client/server*, where there is no possibility of ambiguity.

Abbreviations

If you want to use a shortened technical term like *mux* or *synch* then a final dot to indicate the abbreviation is unnecessary. There is also no need for a dot after titles (Mr Smith, Dr Jones), acronyms (BBC, RDBMS, UK) and times when just the hour is given (9 pm). In the USA, dots tend to be used more often, so follow local standards.

Latin abbreviations, such as *etc.*, *e.g.*, *i.e.* and *viz.* do need dots in order to indicate that the words have been shortened. But put them in the right place – for example, *ad. hoc.* and *e.t.c.* are not correct. If *etc.* is at the end of a sentence, only one dot is needed.

Ellipsis

An ellipsis (three dots) is used to indicate the missing words in quotations that are too long to include in full, as here:

Our Chairman has stated, "This is the year that we will ... really get to grips with the Internet".

The other use of an ellipsis is to signal a conclusion that the readers are supposed to draw for themselves, for example:

> If we cannot get any capacity forecasts by the end of the year then the product launch will be delayed. And if that happens ...

The ellipsis is not a substitute for a dash or semicolon, so the following is incorrect:

> I experimented with a cordless mouse, but sometimes the buttons failed to respond ... maybe radio interference was responsible.

If the ellipsis ends the sentence then you don't need a full stop as well:

> Re-partitioning the disk was our last hope. We waited and prayed ...

PUNCTUATION SUMMARY TABLES

Tables can be downloaded from www.itprojectestimation.com summarising the following punctuation rules:

- Stops
- Apostrophes
- Other marks

Each table is a one-page Microsoft Word document that can be printed out and pinned up in your work area. They can be altered to correspond with local conventions.

BULLETED LISTS

General rules

Bulleted and numbered lists add clarity to a long list of points by introducing white space. Look at this text:

> The requirement for single entry in the devolved systems is far reaching and has serious implications – fundamentally it means that data collection systems must send the data that they collect to the product systems that require the data, they must do so in a form which the receiving (existing system) can handle, and will, in early phases at least, use existing interfaces and must also therefore use existing keys by which the data is known at the receiving system.

It becomes much clearer if expressed with a bulleted list:

> The requirement for single entry in the devolved systems is far-reaching and has serious implications. Fundamentally, it means that data collection systems must:
>
> - Send the data that they collect to the product systems that require it.
> - Do so in a form that the receiving (existing) systems can handle.
> - (In early phases at least), use existing interfaces and, therefore, use existing keys by which the data is known at the receiving systems.

Like all prose techniques, bullets can be over-used. I have reviewed documents where nearly every section consisted of a long list of bullet points, some covering several pages – I saw spots in front of my eyes for the rest of the day. In general, a bulleted list should consist of three to eight points, each of two to five lines. When the points start extending to the quarter-page mark, bullets are no longer appropriate

Try to avoid more than one level of bullet and definitely don't have more than two. If you must have a sub-list, make sure it is short or the overall sense will be lost. A different bullet character is needed for the sub-list; I find that the hyphen or dash is too weak for this and favour the unfilled diamond shape. Whatever you choose, you must ensure that the bullet shapes are consistent throughout the document.

Long lists are better numbered than bulleted; it makes it easier to remember where you are. Otherwise, numbered lists should only be used if the points are in some sort of sequence or order of priority, or if you wish to refer to the individual points later.

Avoid placing a numbered list 'in line' within a sentence, especially if the numbers aren't referred to later. The following example would be better as a bulleted list, or just separated by commas:

> The computer memory is in three sections: (1) program, (2) non-volatile RAM, and (3) scratch-pad RAM.

Finally, remember that unless the bulleted points are in some logical order, the most important or interesting ones should come first.

Grammar in bulleted lists

Each bullet must make complete sense as an extension of the lead-in paragraph, which is not the case here:

> Our proposed plan will benefit MegaCorp & Co. because of:-
>
> - The provision of services, staff and products on advantageous terms,
> - Hackitout Software staff are high quality. They will be able to apply appropriate standards to their work, deliver on time and work with maximum initiative and minimum supervision,

- Filling MegaCorp's perceived shortage of skills at the right time and for the right duration, so reducing costs and
- Our Account Team are motivated to meet the needs of MegaCorp's requirements

As a start point...

In that example, some combinations are gibberish, for example:

Our proposed plan will benefit MegaCorp because of Hackitout Software staff are high quality...

It is easy to proofread each bullet as a continuation of the lead-in paragraph, so there is no excuse for this sort of mistake. Each bullet usually starts with the same type of word, for example:

The API has facilities for:

- Retrieving records with specific key values.
- Retrieving records in sequence.
- Inserting records into the database.
- Deleting records from the database.
- Modifying and replacing records in the database.

If several points begin with the same piece of text, this should be moved into the lead-in paragraph and the remaining points restructured as necessary.

The lead-in sentence must not continue after the list of bullet points because it is too difficult to remember what has gone before, as here:

To achieve these objectives, the company has introduced:

- a new quotations system
- workflow and document imaging
- tele-servicing, as part of a front/back office client-centred approach

to support the processing of customer enquiries and new business.

If possible, all the bullet points should be positive. In the next example, the one positive point seems to inherit the previous "nots":

The selected database should:

- Not lock an entire table every time there is an update.
- Not use a non-standard interface instead of SQL.
- Not require specialist skills for development.
- Provide a higher read-only throughput than at present.

Punctuation and spacing of bulleted lists

The rules are as follows:

- The lead-in paragraph must end with a colon.
- All the bullet points must end with the same type of punctuation mark.
- Avoid mixing complete sentences with fragments. In such cases, each point should be formed into a proper sentence, starting with a capital letter and ending with a full stop.
- If all the points just consist of a word or short phrase then they need no punctuation mark to end them; they can start with a lowercase letter if this is appropriate.
- There must be a blank line before and after the list.
- Usually, there is a blank line between each bullet point. However, if the points are short, or the list is not particularly interesting, then the spacing between the items can be closed up.
- The text following the list should be a new sentence, starting with a capital letter.

The first word or short phrase in each bullet point may be concluded by a full stop, even though it is not a complete sentence. Alternatively, a dash can be used to separate the first word from a short explanation that follows. Sometimes it is effective to put the first word in bold text.

ITALIC, BOLD AND UNDERLINED TEXT

These styles should be used sparingly because they are harder to read than the normal font. Apart from their use in headings, they can be employed as follows:

Italic

- To stress certain words, as in:

The green and yellow wire must *not* be connected to the live pin.

Actually, I get irked by the constant use of italics to stress the word *not*, as in that example and here:

The contract does *not* include any provision for support once the system has been in live use for three months.

Such italicisation may be valid if it comes after a list of things we *will* do, but *not* is such a small word that the italics usually aren't noticed – and what do they add anyway?

- For foreign phrases, such as "The utility can be used for *ad hoc* queries".
- For the titles of referenced documents. You may also use italics for generic documents or those not yet written, as in this example:

  A Domain consists of one or more capabilities and is described in its *Domain Architecture* Document. The constituent capability or capabilities are then further defined in their own *Capability Architecture* documents.

Bold

- For the first sentence, or first few words, of each item in a bulleted list. This summarises the subject of each bullet and so facilitates speed-reading.
- To introduce new terms, as here:

  The next phase in the process is the **Project Justification**, during which a business case is established.

- For unnumbered sub-headings in cases where there would otherwise be an excess of numbered heading levels.
- For important phrases or sentences that summarise your argument. Typically, these will be at the start or end of a paragraph or section.

Underlined

- A nastier alternative to most of the uses of bold and italic. Avoid.

CAPITALISATION

Most of the rules for capitalisation are well known and rarely misused. However, there are a few cases where capitals should be used but sometimes aren't:

1. For headings, titles, captions, table column headings and the like. Usually the first word is capitalised, as are all other words unless they are conjunctions (*and, or*), articles (*a, the*) or short prepositions (*from, on, out, into*).
2. For hardware items and software options, such as a *Power switch* or an *Exit button*.
3. When referring to specific job titles or department names. For example, you would say "Mary Smith, the Senior Buyer for MegaCorp Limited" or "Please send your cheques directly to our Accounts Department".
4. When referring to a named organisation. So you could initially mention *BigCo Bank* but then just refer to the *bank*.

5. When using a term you have defined. For example, I recently wrote a document describing the various phases in a project process: the Project Justification Phase, the Pre-Project Phase and so on. Having defined those terms, I continued to refer to them in capitals. This made it clear that I was talking about the term that I had defined and discussed earlier, not some vague abstraction.

   However, don't overuse this technique and start applying capitals to every specialised term, as this writer has done:

   > All Customer Orders are manually checked before being passed to the Order Entry Supervisor, who allocates them in Order Batches of 24 Forms to a free Order Entry Clerk.

6. When referring to specific sections in your document. It is okay to use lowercase if you are not mentioning the actual section number or title, as in "This section describes the proposed system architecture". However, when referring to a *specific* section it needs the capital, as in "This topic is further discussed in Section 3.2". The same applies to other referenced elements, such as figures, tables, examples, appendices, rules and exercises – but not pages.

7. For the first word in a quotation that forms a complete sentence, even if it is inside another sentence:

   > Our CEO said in 1999, "We must truly embrace the Internet".

   But a capital is not needed when the quotation is a fragment of a sentence, as here:

   > My manger said he was "completely dumbfounded" when I submitted my timesheet early.

NEVER capitalise whole words for emphasis; italic or bold text is better.

Acronyms

Here are some rules for acronyms:

- Don't define one and then only use it once.
- Don't define one (for example, in a glossary) and then not use it.
- Don't use one without defining it, unless you are sure that all your readers will know what it means.
- Don't be inconsistent about the acronym or its expansion.
- Don't repeat words already in the acronym, as in "PIN number" or "ISA account".

Ideally, you should expand the acronym the first time it is mentioned and then only use the acronym:

> All user access will be through the Product Delivery System (PDS). The PDS is a . . .

TABLE 7.1. Example glossary		
Acronym	Expansion	Explanation
PDS	Product Delivery System	Logically separate database system through which all user access is channelled.

However, it may help to redefine the acronym in each section where it is used (for example, once in the Management Summary and again in the proposal itself), otherwise some readers may be confused.

A glossary is useful in cases where a significant number of new acronyms are introduced. It is helpful to provide a short explanation of what each acronym means, as well as defining its expansion; an example is shown in Table 7.1.

THE PROPOSAL LIFECYCLE – REVIEWED

We've not looked at the proposal lifecycle for some time, but we have now finished another stage: the completion of the draft of our proposal, as shown in Figure 7.1.

To create this draft, we needed to follow three sets of rules:

1. The rules for creating tight, targeted, persuasive prose.
2. The rules of grammar.
3. The rules of punctuation.

Having completed your draft, you will have all sorts of good points, and other odd fragments of text, that don't seem to fit anywhere. It seems a shame to lose these perfectly valid, and lovingly crafted, bits and pieces. Worse still, with a jointly authored document you may risk offending someone by omitting some of their musings. But the pressure to stuff all these words into the document somewhere must be resisted. Don't make the same point more than once, don't include waffle that doesn't add to your argument and don't be afraid to throw away any redundant text.

Less is more. Good prose makes its points in the minimum space, in the clearest language. The more you can take out, the more effective the remainder will be. Unfortunately, deleting all the wordiness and repetition is hard work – as Blaise Pascal once stated:

I have made this a long letter because I haven't time to make it shorter.

But such revision is time well spent if the result will sustain the reader's interest and enable your message to be understood.

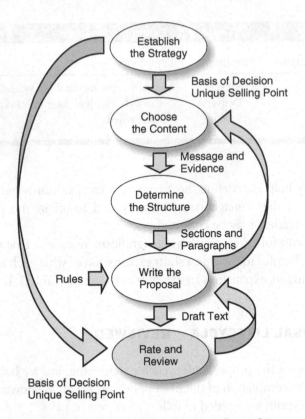

FIGURE 7.1. Proposal lifecycle, stage 4 complete

Before you present your work to others for appraisal, you must undertake a review of the draft within the proposal team. This will tidy up any loose ends, validate compliance with the original request, and ensure that the proposal presents an image of quality and professionalism.

CHAPTER 8

Finishing off

VALIDATING COMPLIANCE

You must do what the customer asks. If you are responding to an Invitation to Tender, Request for Proposal, or similar document, you must check that you have complied with all its requirements. These fall into two groups:

- **The physical requirements**, which may include the expected format of the proposal, a list of sections to be included, how prices are to be presented, how many copies are needed and so on.
- **The technical requirements** defining the system for which you are bidding.

The best method is to draw up a **Compliance Matrix**. On the left-hand side this lists the requirements from the customer, sentence by sentence, in the form of a cross-reference, the text itself, or both. The right-hand side shows how your proposal addresses each requirement. The result will be something like Table 8.1.

As you complete the matrix, you should be asking yourself if your response to each requirement is adequate, and presented in the manner asked for. You must provide a response to all the requirements in the Invitation to Tender, no matter how repetitive or pointless they may seem. So don't just respond with "As before", "See the attached" or "Compliant" – give a full answer to each question asked.

You may choose to include the Compliance Matrix as an appendix to your proposal, either because the customer has asked for one, or because you would like to help them with their evaluation process. In either case, a table consisting of two sets of paragraph numbers may be sufficient, but a fully descriptive matrix explaining exactly how you will meet each requirement is better, if this can be achieved without repeating the entire proposal.

TABLE 8.1. Example compliance matrix			
RFP Reference	Requirement	Proposal reference	Compliance
2.1 (1)	Ease of use	3.2.4 4.1.3	User interface will adhere to Microsoft guidelines. Context-sensitive help will be available on all user screens.
2.1 (2)	Expandability	4.5.3	Main memory is expandable from 128 MB to 512 MB.

PROTECTING YOURSELF

Proposals are expensive to produce and often contain a considerable amount of free analysis. Inevitably, there is a risk that your work will be passed to your rivals or that the customer will take all your good ideas and implement them internally. To protect against this, here are some of the measures you can take:

- Arrange a non-disclosure agreement. Customers sometimes insist on such an agreement in order to protect the information that they supply to you, so it is not unreasonable to ask that your original work be similarly protected.
- Include a proprietary statement in the proposal or covering letter. Typically, this reads something like this:

This proposal contains proprietary information, which must be maintained in confidence. Unless you have obtained written permission, please do not reproduce any information from this proposal, store it electronically, or disclose it to any person not directly responsible for its evaluation.

Note the lack of legalese, although you may like to get your organisation's legal department to define exactly what needs to be said.

- Mark the pages containing particularly sensitive information with a footnote, watermark or similar method.
- Don't go into too much detail. The aim of the proposal is to show you can do the job; it is not the system design. If you can imply that you have analysed the problem to a deeper level than you have revealed, the customer may think it is easier to let you carry on than to establish an internal project. Similarly, do not provide a complete hardware manifest down to the last cable and part number. This just presents the customer with a shopping list – and they may decide to buy elsewhere.

- Don't submit the proposal too early. Usually this is not possible anyway; there is rarely sufficient time to complete it at all. But if the document is floating around for too long, it may be copied or fall into the wrong hands.
- Don't submit more copies than are asked for.

ILLUSTRATING YOUR ARGUMENTS

I am discussing illustrations and tables now, not because I think they should be created after the text of the proposal but because this is the point at which you should be assessing the following factors:

- Do the diagrams actually illustrate the points we are making?
- Do they reveal or imply new information that is not mentioned in the text?
- Can they be simplified?
- Are they referenced within the text in a way that flows logically?
- Are there any other areas where a diagram would help?

Typically, the illustrations are produced in parallel with, or even before, the text itself. Sometimes, this shows – the diagrams don't enhance the arguments, and may not even be mentioned, hanging around like a side dish nobody has ordered.

Illustrations can be useful – they break up the text, make points clear and provide a chance to improve the overall 'professionalism' of the proposal. But they can have the reverse effect, being over-complicated, unnecessary or patronising. In the latter category, I would include clip-art images, such as a handshake when partnership is being discussed, or a moneybag where prices are mentioned. I can manage such concepts without visual aids, thank you.

Some rules for diagrams and tables follow:

- They should be numbered and titled.
- They should be referred to by number, not by such phrases as "the diagram below", so that the illustration can be moved, maybe to a separate page.
- They should be located as close as possible to the place they are referenced, not placed in an appendix.
- They should not be printed at right angles to the text.
- They should have a consistent look. Tables should have a uniform style for headings, borders and content. Diagrams should use the same components and colours.
- They should not be over-complicated. There is nothing wrong with the same illustration appearing more than once, maybe in different variants. For example, a hardware architecture diagram may be used to explain the proposed configuration, then to highlight different data flows, and then to show how much each element will cost. But not all at the same time.

- Tables are often skipped, so should not contain important information not mentioned elsewhere. If this can't be avoided (for example, where a table lists the individual elements of the price) then repeat the important elements as a summary within the main text.

CHECKING THE DETAIL

It's time to get rid of those niggling little errors. Unfortunately, if any text needs to be rewritten following the external reviews then all this must be done again. But we need to present the best copy we have for review, so it's best to get these chores out of the way as soon as possible. Much of the low-level checking can be automated, but this must be augmented by a detailed scrutiny by each member of the proposal-writing team.

Automated checks

A spelling checker will not spot a word that is valid but in an incorrect context. For example, I have a habit of typing *complaint* when I mean *compliant*. No spelling checker is ever going to pick that up and I don't want to put it into the auto-correcting dictionary in case I ever want to use *complaint*. It usually takes another reviewer to spot these sorts of errors.

Don't assume that the spelling checker is wrong and then present a document with some glaring errors. For example, *consensus* and *supersede* really are the right spellings. And remember that it is not our job to invent new words; when the spelling checker rejects *historicisation*, *nonperipheral*, or *disinstigate* (to name just a few of the new coinages I have seen), it is for a good reason.

Many people will take some notice of the red lines flagged by the spelling checker in Microsoft Word but ignore the green lines inserted by the grammar checker. This is foolish – it really is very good. Of course, it sometimes gets confused, but this may be a pointer that your sentence structure is not as clear as it might be. The grammar checker will show up mistakes in many of the areas discussed in this book. And, if any of its little quirks annoy you, just turn off that particular type of check.

Spelling and grammar checking are not all that can be achieved by automated means. You can undertake the following tasks using a "search and replace":

- Standardise capitalisation.
- Remove hyphens where you want dashes.
- Remove unwanted *Notes* and *NBs*.
- Regulate the use of dots in phrases like *e.g.*, *i.e.* and *etc.*

- Standardise the use of terms – for example, if one writer has used *I.B.M.* and another *IBM*.
- Use *we* and *you* rather than company names.
- Remove legalese phrases, such as "herein".
- Remove forbidden phrases, such as "time is of the essence".

Manual checks

Having undertaken all possible automated checks, you will need to undertake a manual scan for the following:

- Adherence to the grammar and punctuation rules.
- Words rightly spelt but in an incorrect context.
- Consistency in terminology (for example, defined terms, names of documents, acronyms and abbreviations) where this can't be done automatically.
- Consistency between lead-in sentences and bulleted lists.
- Punctuation of bulleted lists.
- Capitalisation of words in headings, tables and diagrams.
- Consistency of numbers where they are quoted in different places.
- Accuracy of figures and totals.
- Unnecessary cross-references and footnotes.

It is all too easy to scan the words of a document without actually reading them. I saw a magazine article recently that, in a rather densely worded paragraph, contained the sentence:

> Many new relationships will become apparent as will other currently unseen benefits will arise.

Such mistakes are easily made, but look sloppy. So if you find you are falling asleep and just looking at the words without thinking about what they mean then it's time for a break. Or maybe it's time to revise that section of the proposal in order to make it more stimulating.

CREATING AN IMPRESSION

Now the content is complete, we can turn our attention to the overall look and feel of the proposal. However impressive your arguments, the document must look appealing to read. Something scruffy, closely printed and unillustrated may be ignored or skimmed by some readers, while those that do penetrate further will have started with a negative reaction.

Letter proposals, and those presented internally within an organisation, do not need a great deal of dressing up. In this section, I am more concerned about competitive situations, where your proposal is being evaluated alongside those from your rivals. For all you know, the competing organisations may be as technically capable, may be offering as good a price and may be equally chummy with the customer. But even if the content of your proposal is no better than the rest, at least it can *look* better. A luxury finish is a small investment relative to the cost and time that have been expended on the text itself.

So we need to pay attention to both of the following:

- The internal impression – how the individual pages are laid out and illustrated.
- The external impression – what the overall document looks like.

The internal impression

Some factors to consider here are as follows:

- Use two columns. A multi-column layout is easier to read, allows more formatting options and permits a smaller type size. An asymmetric layout of one-third of the page width for the left column and two-thirds for the right is an effective and flexible format.
- Break up long tracts of text with new paragraphs, bulleted lists and additional section headings.
- Include more tables or diagrams, in accordance with the rules described earlier.
- Don't bury important facts in dull paragraphs or tables. Look for any major points that can be highlighted or boxed, grabbing the attention of a speed-reader. Inset boxes can be headed "Key Point" or suchlike and given a different font or colour.
- Use inset boxes for quotations (for example, from the customer's Invitation to Tender or an industry survey), technical summaries (such as the configuration of a server), facts about your organisation and similar self-contained information.
- Look out for 'widows' (headings on different pages from their text) and 'orphans' (the last line of a paragraph appearing on a page by itself). Similarly, check for bullet lists separated from their lead-in paragraph and orphaned bullet points. There is rarely any harm in starting a new page, even if it leaves some blank space.
- Beware of unfortunate line breaks, as here:

  Hackitout Software has a great deal of experience in creating a disaster
  back-up plan and will recommend and provide a suitable solution ...

- Use plenty of white space, with wide margins.
- Mix different fonts and styles. Use a ten- or twelve-point serif type for the main text.

- Make the spacing between the lines sufficient, so the text is easy to read. The spacing may be varied for different sections of the proposal (for example, double spacing for the management summary and single for the technical descriptions).
- Don't justify the right margin. 'Ragged right' text looks less formal and is easier to read.
- Don't overdo it. A riot of different colours, fonts, graphics and page layouts will detract from your arguments.

The external impression

Factors to consider here include the following:

- Use thermal or 'perfect' binding. Although the resulting document will not lie flat, it will look more professional than one in ring binders, sliders or spiral clips.
- Choose an eye-catching title, descriptive of the benefits the proposed work will bring. For example, don't call it "A Proposal for a Database Strategy" but "Saving Data Entry Costs: The Case for a Distributed Database". This accents the business driver behind the proposal and gives some indication of what is being proposed.
- Use a subtitle to link the name of your organisation with that of your customer – for example, "A Hackitout Software Solution for the MegaCorp Accounts Department".
- Design a front cover. This may include the customer's logo, so long as there are no restrictions on reproducing it. Best of all is a piece of original artwork illustrating the proposal content in some way.
- Use glossy or heavy card for the cover.
- Use good-quality paper, not the standard stuff from your office copier. It must not be possible to see through to the printing on the other side or on the next page.
- Print important sections, such as the Management Summary, on one side of the paper only.
- Use colour, for both the cover and the content, especially the illustrations. But keep it plain and simple – maybe a pale blue for the header and footer, and black for the main text.
- Make it easy to turn to individual sections. If you are using a ring binder then print card dividers with the section titles.
- Don't bind in any pre-printed material. Include it separately or provide a folder to contain all the documents.
- Give everyone an original copy – not a deluxe version for your primary contacts and a photocopy for everyone else. And make sure you provide sufficient original copies so that the customer does not need to create inferior versions internally.

- Make it *different*, by choosing an unusual paper size, coloured paper, pull-out diagrams, a 'facts card' with key points listed – anything that will distinguish your proposal from the pack.
- But, again, don't overdo it. You want a smart and professional document, not something that looks like it was designed by a three-year-old with too many crayons.

CHAPTER 9

Reviewing the result

TACKLING THE REVIEW

The more your proposal is reviewed, the better it will get. This may seem an obvious point, but it is often the case that the review stage is dropped so that the contributors have time to complete their allotted sections. It is far better to present a proposal that is generally correct, coherent and persuasive than one that has every detail nailed down somewhere within a mass of ill-arranged and unchecked paperwork.

Reviews of proposals too often concentrate on the content as written, not how the message can be improved. Of course, we need to ensure that our facts are correct and that we have covered all the risks. But we also need to ensure that the arguments we are making, such as "Why buy from us?" and "Why should you take the action I recommend?" are being expressed in a way that will get results. While reading the examples in this book, I'm sure many of the mistakes jumped out at you. You can recognise a badly made point or a paragraph not aimed at the correct audience as easily as spotting a spelling error. If you apply that mindset the next time you review a proposal then you will be able to tune the text to its maximum effectiveness.

I have previously mentioned the benefit of mentally reading the proposal 'aloud'. Try to imagine what each potential reader would think and the questions he or she might ask. The more you can enter the character of each reader, the better you will be able to spin your arguments around their particular personalities, circumstances and problems.

REVIEW METHODOLOGY

Reviewers may or may not be familiar with the problem being addressed by the proposal. Ideally, you want a mixture of those who understand the situation, or who have read the Invitation to Tender, and those who are seeing it all for the first time. The latter group will be best able to comment on the clarity and persuasiveness of the arguments in the proposal – if you can convince *them*, your customers should be easy.

Beware of last-minute changes. Text that is shoved in after the proposal has been thoroughly scanned and cleansed is often error prone and may even destroy the overall structure and flow. In particular, don't allow senior staff to revise the 'live' document. They don't have the time to study the structure, they insert what they think are meaty points in all the wrong places and they remove important text without telling anyone. Instead, tell all reviewers to submit their revisions as comments then consider each one in the context of the overall structure.

Careful review of a proposal needs time, so will cost money in terms of the resources needed. However, proposals are the lifeblood of your business and there is no point in trying to economise on this final step. The time spent in reviews must be proportional to the amount of effort that has been devoted to the preparation of the proposal in the first place. The reviewers must be pre-warned about how much effort is needed, and then given sufficient time to complete the task.

So who should review the proposal? I would suggest some or all of the following:

- **The proposal team** – which may include the Bid Manager, estimators, technical experts, co-ordinators and so on. Each member will have devoted most of the effort to their own responsibilities or sections, but now they need to take a step back and analyse the document as a whole.
- **Quality assurance staff** – you may have a separate QA department, or you may nominate someone, such as the keeper of the Document Standard discussed in Appendix A.
- **Line management** – in particular those who will be signing off the proposal before it is issued.
- **External people** – who have not been involved with the project until this time. They may have a specific technical interest or a general level of experience.

For important bids, you should consider establishing a 'Red Team'. This group remains disconnected from the main Bid Team until a draft of the proposal has been produced. They use the intervening time to become thoroughly conversant with the ITT, but do not attempt to devise a solution. When the draft proposal is ready, they review it from the customer's perspective and so are able to provide constructive feedback about the validity of the solution, the persuasiveness of the text, the strength of the Unique Selling Points and so on. They then work together with the original Bid Team to strengthen the proposal and eliminate their criticisms.

THE PROPOSAL EVALUATION QUESTIONNAIRE

Each reviewer needs to understand what to look for and to apply a consistent set of assessment rules. For this reason, I have devised the **Proposal Evaluation Questionnaire**, an Excel workbook that can be downloaded free of charge from www.itprojectestimation.com.

Instructions for use

Each sheet in the workbook can be selected from the tabs at the bottom of the screen. The "Instructions" tab repeats the information in this section. The "Header" tab contains basic information about the proposal and the reviewer. It also shows the final rating once the questionnaire has been completed. The remainder of the questionnaire is set out as six sets of "Assertions" that cover various aspects to the proposal under review, as follows:

- **Empathy** – the current situation and the Basis of Decision.
- **Content** – the substance of individual sections, such as the Management Summary, technical descriptions, commercial terms and appendices.
- **Persuasiveness** – the Unique Selling Point and the development of arguments.
- **Clarity** – simplicity of language, an appropriate tone and avoidance of waffle.
- **Accuracy** – spelling, grammar, numbers and compliance.
- **Appearance** – the overall look and feel, illustrations and readability.

Figure 9.1 shows the start of one of the sets of Assertions.

Empathy	Strongly Disagree / Not Applicable	Disagree	Uncertain	Agree	Strongly Agree	
Assertion						**Notes & Comments**
We show we understand the business problems to be solved.	x					No mention in Section 1
The customer's Basis of Decision is stated and clear.		x				Briefly hinted at in Section 1
We address all the points in the Basis of Decision.		x				
Previous attempts to solve the problem are not criticised.				x		
We show we have an emotional commitment to solving the problem.			x			Mainly techncial detail
We show we can be relied on to solve the problem.				x		Good references
We evoke emotions in the reader (e.g. inspiration, encouragement, comfort, shock, stimulation, and worry).		x				Should try to do this more in Section 1.7

FIGURE 9.1. Questionnaire assertions

For each Assertion, the proposal should be rated, by selecting one of the six possible options, as follows:

- **Inapplicable** – the Assertion is not relevant to this proposal.
- **Strongly disagree** – the Assertion is completely untrue.
- **Disagree** – the Assertion is generally untrue.
- **Uncertain** – the validity of the Assertion is too hard to determine.
- **Agree** – the Assertion is generally true.
- **Strongly agree** – the Assertion is very true.

If an Assertion is not given a rating, is given more than one rating, or is not selected with an "x", then the ratings cells are highlighted in red. Results will not be accurate if such ratings are not corrected.

A red triangle in the corner of a cell shows that there is some help or other note appended. Hold the cursor over the cell to see the text. Owing to an undocumented feature in Microsoft Excel, the size and position of these notes are somewhat unpredictable. If the note is not completely visible, right-click the cell, select "Edit Comment" and then manipulate the notes window until the text is fully visible. If you no longer wish to see the notes, select "Tools", "Options", "View" and then choose "None" under "Comments".

If you wish to add a comment to your rating, note the relevant proposal section numbers in the space provided or on a separate sheet.

The workbook may be saved under another name and then restored without affecting any ratings already made.

Don't be scared to allot the extreme ratings of "Strongly Agree" or "Strongly Disagree". Your business depends on getting proposals right, so the aim of the review is to tell the truth, not to avoid offending the authors. If something is good then say so. If it is bad then the authors will want to know.

Once all the questions have been answered, the "Header" sheet depicts the rating for each category, as well as an overall rating, as shown in Figure 9.2.

What is an acceptable score? That's up to you, but I don't allow any of the values to drop below 75% for proposals issued to my customers or management.

Modifications to the questionnaire

Under the "Instructions" tab in the workbook are directions for the following maintenance actions:

- Removing Assertions.
- Adding new Assertions.

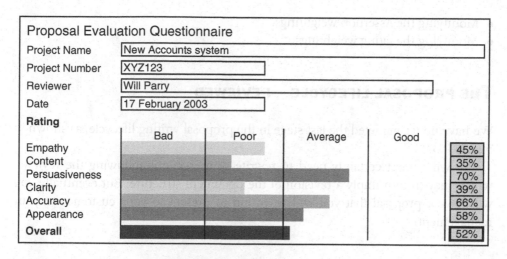

FIGURE 9.2. Questionnaire ratings

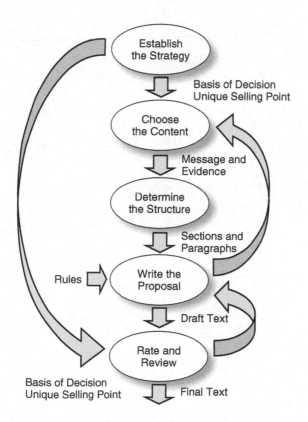

FIGURE 9.3. Proposal lifecycle complete

- Modifying the Assertion weightings.
- Modifying the rating weightings.

THE PROPOSAL LIFECYCLE – REVIEWED

We have now completed the last stage in the proposal writing lifecycle, as shown in Figure 9.3.

We will almost certainly need to rewrite some sections following the reviews, which may in turn imply a revision of the document structure. But eventually you will have a proposal that you will be proud to present to your customer or your management.

CHAPTER 10

Summary

Table 10.1 shows the twelve rules for creating a winning proposal. Regrettably, I cannot guarantee that applying these rules will win that big contract, or that your pet project will be approved. Maybe the technical solution is inappropriate or too expensive – no amount of word-spinning can disguise that. But I can promise that if you follow these rules then your proposal will be as persuasive as can be achieved in the time available. No one can ask for more than that.

TABLE 10.1. Twelve rules for a successful proposal

1 List the Basis of Decision and the Unique Selling Point – remember that *every word* in the proposal must address these.

2 Decide on your strategy – what is the major theme?

3 Determine the content of the proposal before the structure – there is no universal template.

4 Plan the content *before* you start to write.

5 Structure every section and paragraph as a triangle, using headings as mini-summaries.

6 Avoid a dry, abstracted tone – try to raise emotions in the reader.

7 Spin and tune each sentence – make every word count.

8 Check that you have met all the physical and technical requirements mandated by the customer.

9 Get as many people as possible to review the document.

10 Make sure there are no stupid mistakes in spelling, grammar, punctuation or use of words.

11 Tell reviewers to comment not just on the content but on the way that content is presented – use the Proposal Evaluation Questionnaire.

12 Make your proposal *different* – invest in quality printing and presentation.

Appendix A
The Document Standard

Why have a standard?

You may have disagreed with some of the rules defined in this book. Matters such as the punctuation of bulleted lists, the use of commas and the acceptability of Latin abbreviations are not universally standardised, and the conventions for punctuation and spelling vary in the different English-speaking countries. You may also have noticed that some of the text in this book does not adhere to the rules I have defined. This is because the Cambridge University Press has its own guidelines, which override my personal preferences. What matters is consistency. Books issued by the Cambridge University Press must be produced to high, uniform standards, which supplant the individual conventions used by each author and outweigh any arguments about whose convention is 'right'.

Every organisation – not just publishers like the Cambridge University Press – needs a standard that defines a 'house style' for the appearance and conventions of the documents it issues. Such a standard is often called a "Style Guide". I dislike this term. Firstly, it is not a "guide"; it is a mandatory standard – no exceptions. Secondly, it is not about "style", which to me is more about the way words are used than about conventions for terminology, punctuation and so on. So I call it a "Document Standard", which makes its function clear.

Usually, there will be a different standard for each type of document that your organisation may produce: letters, technical documentation, user manuals and so on. There will often be a large amount of commonality between these, plus additional conventions that apply to each document type.

Contents of the standard

A Document Standard has two parts. Firstly, there are elements that can be automated through the word processor, such as the page layout and the use of fonts. Secondly, there are elements that must be verified manually, either by "search and replace" operations or by visual checks.

Word processor standards

In Microsoft Word, the Document Standard is usually implemented as a "template", which can be used to define the following:

- **Page layout** – size, margins, headers and footers (font, colour, lines, logo etc.), and security measures (watermark or copyright symbol).
- **Fonts** – the standard fonts can be embedded into the template so they are displayed and printed correctly for all users.
- **Text formatting styles** – for different levels of heading, tables, illustration titles, bulleted and numbered lists, and so on.
- **AutoText** – for boilerplate sections, such as your standard terms and conditions for contracts.

Further standardisation can be obtained by having a common "Custom Dictionary". This holds words that are not to be found in the standard dictionary used by the spelling checker in Word, but which are valid for you or your organisation. For example, I may put "MegaCorp" and "Hackitout" into my Custom Dictionary; this would ensure that they are always spelled correctly and in a consistent format. Terms that are red-lined by Word but which you find acceptable – such as *cashflow* and *costed* – can also be included in the Custom Dictionary,

The common Custom Dictionary can be held as a shared file, augmented by local Custom Dictionaries for each individual writer or for a specific document. For example, if a proposal is to BigCo then that spelling can be placed into a Custom Dictionary so that variants such as Bigco or BigCO will be flagged as incorrect.

Your standard may also define how the options for the spelling and grammar checking facilities are to be configured. Among the grammar elements that may be checked are the following:

- The number of spaces between sentences.
- Whether there should be a comma before the last item in a list.
- Whether punctuation should be within or outside quote marks.
- Whether sentences beginning with conjunctions are permitted.
- Whether use of the first person is permitted.
- Whether split infinitives are permitted.

Manually checked standards

The lists of manual checks can be defined as "hidden text", which allows the user to see the information when they desire, but to eliminate it from the final draft. The standard may include rules and conventions for the following:

- Grammar
- Punctuation
- Style
- Tone
- Date formats
- Abbreviations
- Latin abbreviations (*e.g.*, *i.e.*, *etc.*)
- Numerals
- Capitals
- *Notes* and *NBs*
- Specific terms or company names
- Banned phrases, such as "time is of the essence"
- Cross-references and footnotes

Depending on the type of document, and your own type of business, you may also include standards for the following:

- Code samples
- Screen shots
- Graphical user interfaces
- User procedures
- Internationalisation
- Legal guidelines (e.g. copyright, confidentiality and the use of trademarks)
- Illustrations (line art, graphics, photographs etc.)
- Indexing

Maintenance of the standard

The Document Standard needs an owner, whose responsibilities are as follows:

- Distributing the latest version of the Document Standard to everyone who may need it.
- Maintaining the Custom Dictionary.
- Modifying the Document Standard following user suggestions.
- Reviewing documents to check adherence to the standard.

Appendix B
Select bibliography and resources

This section is kept up to date on www.itprojectestimation.com.

Proposal writing guides

Kantin, Bob. *Sales Proposal Kit for Dummies*. John Wiley, ISBN 0 7645 5375 5.

McCann, Deiric. *Winning Business Proposals*. Oak Tree Press, ISBN 1 86076 166 6.

Porter-Roth, Bud. *Proposal Development: How to Respond and Win the Bid*. Oasis Press, ISBN 1 55571 431 5.

Sant, Tom. *Persuasive Business Proposals: Writing to Win Customers, Clients and Contracts*. American Management Association, ISBN 0 8144 5100 4.

Technical style guides

Blake, Gary and Bly, Robert W. *The Elements of Technical Writing*. Longman, ISBN 0 02 013085 6.

Manual of Style for Technical Publications, Microsoft Press, ISBN 0 7356 1746 5.

Sun Technical. *Read ME First! A Style Guide for the Computer Industry*. Prentice-Hall, ISBN 0 13 142899 3.

English grammar and style guides

Aitchison, James. *The Cassell Guide to Written English*. Cassell, ISBN 0 304 34963 1.

Cutts, Martin. *The Quick Reference Plain English Guide*. Oxford University Press, ISBN 0 19 866243 2.

Fowler, H. W., revised by Sir Ernest Gowers, *Fowler's Modern English Usage*. Oxford University Press, ISBN 0 19 281389 7.

Freedman, Lawrence H. and Bacon, Terry R. *Style Guide*. Franklin Quest, ISBN 0 933427 00 X.

Gowers, Sir Ernest, revised by Sir Bruce Fraser, *The Complete Plain Words*. Pelican Books, ISBN 0 14 020554 3.

Kane, Thomas S. *The New Oxford Guide to Writing*. Oxford University Press, ISBN 0 19 509059 4.

Lindsell-Roberts, Sheryl. *Business Writing for Dummies*. John Wiley, ISBN 0 7645 5134 5.

Manser, Martin and Curtis, Stephen. *The Penguin Writer's Manual*. Penguin Books, ISBN 0 14 051489 9.

Strunk, William Jr and White, E. B. *The Elements of Style*. Allyn and Bacon, ISBN 0 205 30902 X.

Trask, R. L. *The Penguin Guide to Punctuation*. Penguin Books, ISBN 0 14 051366 3.

Other relevant books

Adams, Scott. *Dilbert and the Way of the Weasel*. Boxtree, ISBN 0 7522 6503 2.

Coombs, Paul. *IT Project Estimation: A Practical Guide to the Costing of Software*. Cambridge University Press, ISBN 0 521 53285 X.

Porter-Roth, Bud. *Request for Proposal: A Guide to Effective RFP Development*. Addison-Wesley, ISBN 0 201 77575 1.

Reifer, Donald J. *Making the Software Business Case: Improvement by the Numbers*. Addison-Wesley, ISBN 0 201 72887 7.

Thouless, Robert H. *Straight and Crooked Thinking*. Macmillan, ISBN 0 330 24127 3.

Websites

The Plain English Campaign – www.plainenglish.co.uk.

Services

Eposal (www.eposal.com) provide a method of creating and delivering your proposal online.

IT Project Estimation Limited (www.itprojectestimation.com) provides software and services to help with IT project estimation, costing, proposals and the bid process.

Porter-Roth Associates (www.rfphandbook.com) provides RFP writing services, assists in developing management and technical requirements, provides RFP training sessions, and evaluates proposals.

Pragmatech Services (www.pragmatech.com) market RFP response, proposal automation, presentation generation and content management software.

PresentationPro (www.presentationpro.com) offer resources for PowerPoint presentations, such as themes, backgrounds, photos and video clips.

ROI4Sales (www.roi4sales.com) build Return On Investment (ROI) tools that integrate into the sales process.

SalesProposals.com (www.salesproposals.com) is a consulting organisation that works with companies to design sales proposal models and supporting sales tools.

Sant Corporation (www.santcorp.com) provides software and services to help with sales letters, RFP responses, presentations and proposals.

Appendix C
Case studies

Case study 1

The following proposal is to be made more effective. What steps would you take first and how would you go about rewriting the text to get a positive result? What would be your version of the proposal?

Forward Buying Plan

Companies today face challenges on two fronts from the business environment, with the likes of customer servicing, cost of service, new competition, and new legislation, and from the IT environment, with the likes of Year 2000, European Monetary Union, Internet and Intranet technologies; significant strains can be put on your resources and will require detailed planning and co-ordination.

To achieve success in all these challenges companies need a clear visible, though flexible overall plan for the most appropriate implementation strategies. Project planning including timescales, and access to key skills and experienced resources when required will be essential.

In the current climate resourcing must be planned well in advance to ensure availability. Hackitout Software's "*Forward Buying Plan*" works like a partnership to provide appropriate resources as and when needed. To do so Hackitout Software need to have a continuing view of the likely forward workloads and skill sets needed.

As a start point for establishing this view, Hackitout Software would arrange a joint planning session between MegaCorp and a principal consultant from Hackitout Software. The workshop will identify and scope the business objectives / requirements, and map on to this your outline IT strategy. This exercise will identify individual programmes. The

---

work-shop programme will be free of charge to MegaCorp. Once the programmes are identified both parties can agree:

- priorities
- critical resources
- benefits and risks,
- timescales
- overall project and implementation plans.

Following the work-shop, Hackitout Software would be in a position to construct a purpose built "*Forward Buying Contract*" and/or take on specific items of work as individual projects, whichever is most appropriate in each case.

Typical Forward Buying contracts normally contain some of the following:

- Preferential rates for volume business.
- An agreed framework for all work that Hackitout Software conducts for MegaCorp.
- Preferred supplier status.
- Projects developed on a Time and Materials, Fixed Price, or Risk Share (share profits when project is under budget; share costs if project is over budget) basis
- Supply managed resources (pools of staff with particular skills matched to MegaCorp's needs)
- assign staff to act in an Interim Management capacity if MegaCorp managers are temporarily unavailable
- Application Management.
- Supply certain Products suitable for MegaCorp's business.

Hackitout Software would also assign an experienced senior consultant to work with you to help with the on-going project planning and implementation process. This consultant, will be your principle point of contact, and would be able to help with the project management and monitoring so that Hackitout Software will be able to commit to provide MegaCorp with the right resources at the right time.

Finally the benefits to MegaCorp of using Hackitout Software's services under the "*Forward Buying Plan*" are:

- The provision by Hackitout Software of services, staff and projects on advantageous financial terms. MegaCorp would get preferential rates.
- Hackitout Software staff are high quality. They will be able to apply appropriate standards to their work, deliver on time, and work with maximum initiative and minimum supervision.
- Hackitout Software can operate an Account Team who are motivated to meet the needs of MegaCorp.
- MegaCorp can have the knowledge that their IT requirements will be fulfilled in the future, particularly with the perceived shortage of skills which will affect the market in the coming months.

Firstly, let us look at the Basis of Decision. Here we have a 'cold call' situation, where the reader, presumably someone who is responsible for managing IT projects, is being offered a service. Let us assume that they have heard of Hackitout Software before, so we do not need to introduce ourselves. Our reader will be asking:

- What is this all about?
- Do I have the sort of problem being described?
- Does Hackitout Software's solution make any sense?
- What exactly are they offering?
- What is in it for me?
- What do I have to do next?

So what is our Unique Selling Point? Interpolating some things that are not very explicit in the given text, I think that we are saying:

- We have people who can plug the resource gaps in your IT programmes.
- Our staff are flexible, experienced and can cover the full range of project roles.
- We will offer discounted rates to you.
- We will offer you a *free* forward planning service.
- We can back up this offer with other Hackitout Software services.

It is likely that all of our rivals could offer a similar service, so the thrust of this new initiative has to centre on the fourth of these points – the truly *unique* offering. We are trying to win some business by offering a planning review for free, hoping that this will expose some resource deficiencies at MegaCorp in the short term, and showing that we can plan effectively for the long term.

Looking at the current document structure, it is not really too bad. It starts with the general context and covers the specific problems of resource management. It then describes our proposed service. Things then get a little muddled because the next section covers the contents of a possible service contract before the text returns to a list of the benefits of our proposal.

The structure roughly conforms to the SOAP (Situation, Objective, Appraisal, Proposal) model described in Chapter 4 (Structuring the Proposal). I think this is the right plan, but I would modify it by removing the "Appraisal" section. We do not want to propose a list of alternatives because this would decrease the impact of our assertion that we *know* what the reader's organisation needs. And, anyway, we don't yet know much about their specific problems. Instead of the "Appraisal", I would add a section at the end explaining what the reader should do next – a serious omission from the original text.

After some thought, my sketch-plan for the rewritten proposal is as follows:

- **Situation**:
  - Industry-wide problems.
  - Some companies manage better than yours.

- **Objective**:
  - Plug your gaps with our staff.
  - Plan ahead.
- **Proposal**:
  - We help you create a long-term plan.
  - We have resources and backup services.
  - Initially it is free.
- **First step**:
  - Purpose of the initial planning workshop.
  - Contact us if interested.

Looking back at the Basis of Decision, we can see that this structure will fit fairly well. The reader will understand the problem we are trying to address, what we are suggesting and what their next action should be. We must take care that the solution we are proposing is explicit and that it relates to the problem being addressed.

A review of our Unique Selling Point reveals that the "unique" part – the *free* consultancy – has been rather buried. It must be stressed up front ("start with an earthquake"). The other good things will emerge as we describe our proposal and its benefits. I also want to introduce a note of worry for the readers – making them feel that everybody else is utilising services like the one we are promoting. Remember the "FUD factor": Fear, Uncertainty and Doubt. If we can create such faint feelings of unease, we are sowing the seeds for a request for help where no such desire existed before.

So now we have the answers to the five Pre-write Checks:

- **WHO are the readers?** IT managers who are aware of Hackitout Software's services and who may have a problem in controlling resources.
- **WHAT will influence their decision?** Whether they have a resource management problem and whether our solution seems credible.
- **WHY is our offering unique?** Some free consultancy to plan resources in the short term.
- **HOW will the document be structured?** Describe the current situation, then the objectives, then our proposal and then what the user needs to do.
- **WHERE will I put each major point?** You can see what I decided in my model answer below.

Having decided on our strategy for the document, we come to the tactical decisions of the words to employ and the layout to adopt. The following thoughts occurred to me when I first read the original:

- The title is weak: it refers to a term that is only defined later. And it shouldn't end with a full stop.
- "With the likes of" is not good business English.

- The first paragraph is one long rambling sentence.
- The new term "Forward Buying Plan" doesn't need to be in quotes, italic *and* bold.
- The flow is impeded by poor use of punctuation and incorrect sentence constructions. For example:

Hackitout Software's "Forward Buying Plan" works like a partnership to provide appropriate resources as and when needed. To do so Hackitout Software need to have a continuing view of the likely forward workloads and skill sets needed.

"To do so" is not a good way to begin the second sentence because it refers to the subsidiary action of the previous sentence, "provide appropriate resources", not the primary action, "works like a partnership". A subtle mistake, but it brings the flow to a crashing halt while the reader tries to work out what the "so" is referring to.

- "Workshop" is sometimes hyphenated, sometimes not.
- The whole thing is too long; the same points seem to be made more than once (for example, preferential rates).
- Bulleted lists are inconsistently punctuated and are too widely spaced for such a short letter.
- It is unclear whether this is a generic proposal or one specific to MegaCorp; the company name does not appear until quite late on.
- The section covering the "Forward Buying Contract" doesn't really contribute to the argument and makes the document too long. Such details need not be discussed unless the reader 'takes the bait' and decides to discuss our proposal in more detail. So the good points from this section should be subsumed into the remaining text; the rest can be axed.
- The words about how Hackitout staff apply standards, deliver on time, and work with maximum initiative and minimum supervision are 'motherhood and apple pie' – clichéd and unnecessary if the reader is already familiar with the company.

With all this in mind, you should be eager to start on your rewrite of the proposal. My version follows:

Free Resource Planning Workshop

MegaCorp, like all major organisations, faces resource management challenges. Some – such as customer service, internal costs, competition and new legislation – arise as normal business pressures. But others are generated within the IT environment. Management of Year 2000, European Monetary Union, Internet and Intranet programmes all adds to the strains on resources already stretched by the maintenance and improvement of existing systems.

  Companies who rise to these challenges do so with a clear, visible and flexible overall resource plan – one that matches potential requirements to the most appropriate internal and external personnel. With such a plan, key skills can be applied at the right time and for the appropriate duration, thus minimising costs and preventing delays caused by lack of resource. This is particularly relevant given the shortage of skills that will affect the market in the coming months.

Hackitout Software's *Forward Buying Plan* will provide a clear vision of the resources needed for MegaCorp to succeed with its many parallel initiatives. To meet the demands of the plan in the most cost-effective way, our in-depth pool of skills and experience can be applied to augment the staff already available. We will work in partnership with you to provide the appropriate resources as and when needed.

The benefits to MegaCorp of our *Forward Buying Plan* are:

- Assurance that your IT resource requirements, including those for management staff, can always be fulfilled.
- Preferential rates – from the services of a single individual to the implementation of an entire project.
- Constant availability of high-quality staff with the precise mix of skills and experience you need.
- Access to our industry and technical knowledge, and to our product selection and purchasing capabilities.

We have formed a dedicated team who are motivated to meet the specific resourcing needs of MegaCorp. The team is led by an experienced senior consultant who will act as your principal point of contact, assist with project management and planning, and ensure that the most appropriate Hackitout Software staff are available when needed.

As a first step towards establishing the *Forward Buying Plan* for MegaCorp, we will need to assess your current workloads and the required skill sets. To establish this initial view, we will arrange a *free* joint planning session between yourselves and the leader of our resourcing team. This workshop session will identify and scope the business objectives and requirements of all your ongoing programmes, and map these to your current resource capacity. The aim will be to examine and consolidate the following for each current and planned initiative:

- project and implementation plans
- benefits and risks
- priorities
- timescales
- skill sets
- critical resources
- potential shortages

If you wish to take advantage of this *free* resource planning workshop and create your *Forward Buying Plan*, please contact John Smith at Hackitout Software on 0123 4567.

This is far from perfect, partly because I have tried to preserve some traceability to the original in order to demonstrate how the text has been improved. However, the new version is shorter and more punchy, appeals to the reader's Basis of Decision, brings out our Unique Selling Point, invokes the FUD factor, and leaves no doubt about what to do next.

Case study 2

Here is another letter proposal that needs more punch. This one was issued exactly as shown below – only the names have been changed. I have not provided a model solution, so see how you get on.

Dear Mr Jones

Intro

On 25 Feb we received your Request for Information (RFI) for the Virtual Call Centre. Enclosed you find our response which we hope will lead to continued discussions and ultimately a fruitful relationship.

Our understanding of your needs:

Following current market trends and a clear requirement from MegaCorp's customers, MegaCorp has identified the need the enhance it's current network with value added services. Enhancements to the network such as Intelligent Routing, Network CTI and Network IVR are seen as a first step to building a sustainable competitive advantage. In order to market and roll out these added features on your network MegaCorp is looking for a partner(s) who can assist with the implementation and has the capabitity to execute different business models.

Decision:

We have decided not to respond formally to the RFI but instead re-inforce Hackitout's product independence and capability to integrate which was highlighted in the meeting held on our premises on 22nd September. In addition Hackitout has extensive consulting capability in the Telecoms area. In it's thirty years of operation Hackitout has been widely recognised as a trusted partner who is capable of managing third parties, delivering within budget and on time.

Justification:

Product selection is a task often performed by the customer who will then start further discussions and negotiations with the short listed vendors and their potential imple-mentation partners. Hackitout has build up strategic as well as tactical relationships with many product vendors who have used Hackitout for it's core business; consulting, implementation and system integration expertise. The RFI for the Virtual Call Centre is centered around a significant amount of so called point product solutions such as intelligent routing, network CTI, IVR, workforce management, recording, campaign management , call blending etc. Although we have had an introductory meeting, we feel that Hackitout has, at this stage, insufficient insight into your business to make an informed and business led decision.

Way forward:

We would therefore rather become involved once you have made your final short list and become a advisory partner in order to aid your final selection. At this particular stage Hackitout's value would come into real effect when integration issues, partner selection, partner management, project management, executing business models and risk management become critical issues.

Hackitout Value (refer to White Paper):

In order to proof our commitment and dedication to developing MegaCorp's service offerings into the market, we have enclosed a White Paper. This paper reflects on the carrier industry and our thoughts on how MegaCorp could create real sustainable advantage in a market where margins on network revenue will continue to be under pressure.

Conclusion:

Considering all of the above and the fact that MegaCorp has already invested significant amounts of time in evaluating different product scenarios, we feel that Hackitout should focus on it's core competencies as indicated above. In projects such as these, risk reduction and management is paramount . . . . Obviously, we can provide references and site visits if required and we look forward to developing mutually beneficial relationships with yourself and your customers.

I will contact you on 7 March and discuss the above in more detail.

Yours sincerely

John Smith

Account Manager, Solutions, Telecoms

Index

Printed in the United States
By Bookmasters